Lecture Notes in Mathematics

Edited by A. Dold and B. Eckmann

749

V. Girault
P.-A. Raviart

Finite Element Approximation of the Navier-Stokes Equations

Revised Reprint of the First Edition

Springer-Verlag
Berlin Heidelberg New York 1981

Authors
Vivette Girault
Pierre-Arnaud Raviart

Analyse Numérique
Tour 55–65, 5ème étage
Université Pierre et Marie Curie
4, Place Jussieu
F-75230 Paris Cedex 05

AMS Subject Classifications (1970): 35Q10, 65-06, 65MXX, 65M15, 65N30, 76-XX

ISBN 3-540-09557-8 Springer-Verlag Berlin Heidelberg New York
ISBN 0-387-09557-8 Springer-Verlag New York Heidelberg Berlin

Library of Congress Cataloging in Publication Data
Girault, Vivette, 1943-
Finite element approximation of the Navier-Stokes equations.
(Lecture notes in mathematics; 749)
Bibliography: p.
Includes index.
1. Viscous flow. 2. Navier-Stokes equations-- Numerical solutions. 3. Finite element method. I. Raviart, P. A., 1939-- joint author. II. Title. III. Series: Lecture notes in mathematics (Berlin); 749.
QA3.L28 no. 749 [QA929] 510'.8s [519.4] 79-21917
ISBN 0-387-09557-8

Printing and binding: Beltz Offsetdruck, Hemsbach/Bergstr.
2141/3140-543210

INTRODUCTION

The contents of this publication have been taught at the University Pierre & Marie Curie as a graduate course in numerical analysis during the academic year 1977-78.

In the last few years, many engineers and mathematicians have concentrated their efforts on the numerical solution of the Navier-Stokes equations by finite element methods. The purpose of this series of lectures is to provide a fairly comprehensive treatment of the most recent mathematical developments in that field. It is not intended to give an exhaustive treatment of all finite element methods available for solving the Navier-Stokes equations. But instead, it places a great emphasis on the finite element methods of mixed type which play a fundamental part nowadays in numerical hydrodynamics. Consequently, these lecture notes can also be viewed as an introduction to the mixed finite element theory.

We have tried as much as possible to make this text self-contained. In this respect, we have recalled a number of theoretical results on the pure mathematical aspect of the Navier-Stokes problem and we have frequently referred to the recent book by R. Temam [44]. The reader will find in this reference further mathematical material.

Besides R. Temam, the authors are gratefully indebted to M. Crouzeix for many helpful discussions and for providing original proofs of a number of theorems.

CONTENTS

CHAPTER I

MATHEMATICAL FOUNDATION OF THE STOKES PROBLEM

§ 1 - GENERALITIES ON SOME ELLIPTIC BOUNDARY VALUE PROBLEMS

In this paragraph we study briefly the Dirichlet's and Neumann's problems for the harmonic and biharmonic operators.

1.1. Basic concepts on Sobolev spaces

Our purpose here is to recall the main notions and results, concerning the classical Sobolev spaces, which we shall use later on. Most results are stated without proof. The reader will find more details in the references listed at the end of this text .

To simplify the discussion, we shall work from now on with real-valued functions, but of course every result stated here will carry on to complex-valued functions.

Let Ω denote an open subset of $\mathbb{R}^n$ with boundary Γ . We define $\mathcal{D}(\Omega)$ to be the linear space of functions infinitely differentiable and with compact support on Ω . Then, we set

$$\mathcal{D}(\bar{\Omega}) = \{\varphi_{|\Omega} \;;\; \varphi \in \mathcal{D}(\mathbb{R}^n)\} \quad ,$$

or equivalently, if $\mathcal{O}$ denotes any open subset of $\mathbb{R}^n$ such that $\bar{\Omega} \subset \mathcal{O}$,

$$\mathcal{D}(\bar{\Omega}) = \{\varphi_{|\Omega} \;;\; \varphi \in \mathcal{D}(\mathcal{O})\} \; .$$

Now, let $\mathcal{D}'(\Omega)$ denote the dual space of $\mathcal{D}(\Omega)$, often called the space of distributions on Ω . We denote by $< . , . >$ the duality between $\mathcal{D}'(\Omega)$ and $\mathcal{D}(\Omega)$ and we remark that when f is a locally integrable function, then f can be identified with a distribution by

$$< f,\varphi > = \int_\Omega f(x)\varphi(x)\,dx \qquad \forall\, \varphi \in \mathcal{D}(\Omega) \; .$$

Now, we can define the derivatives of distributions. Let $\alpha = (\alpha_1 , \ldots , \alpha_n) \in \mathbb{N}^n$

and $|\alpha| = \sum_{i=1}^{n} \alpha_i$. For u in $\mathcal{D}'(\Omega)$, we define $\partial^\alpha u$ in $\mathcal{D}'(\Omega)$ by :

$$< \partial^\alpha u,\varphi > = (-1)^{|\alpha|} < u,\partial^\alpha \varphi > \quad \forall \varphi \in \mathcal{D}(\Omega) ;$$

i.e. if $u \in \mathcal{C}^{|\alpha|}(\bar{\Omega})$ then $\partial^\alpha u = \dfrac{\partial^{|\alpha|} u}{\partial x_1^{\alpha_1} \dots \partial x_n^{\alpha_n}}$.

For $m \in \mathbb{N}$ and $p \in \mathbb{R}$ with $1 \leq p \leq \infty$, we define the Sobolev space :

$$W^{m,p}(\Omega) = \{v \in L^p(\Omega) \ ; \ \partial^\alpha v \in L^p(\Omega) \quad , \quad \forall \ |\alpha| \leq m\} ,$$

which is a Banach space for the norm

$$(1.1) \qquad \|u\|_{m,p,\Omega} = \Big(\sum_{|\alpha|\leq m} \int_\Omega |\partial^\alpha u(x)|^p dx\Big)^{1/p} \quad , \quad p < \infty$$

or

$$\|u\|_{m,\infty,\Omega} = \sup_{|\alpha|\leq m} \ (\operatorname*{sup\,ess}_{x \in \Omega} \ |\partial^\alpha u(x)|) \quad , \quad p = \infty .$$

We also provide $W^{m,p}(\Omega)$ with the following seminorm

$$(1.2) \qquad |u|_{m,p,\Omega} = \Big(\sum_{|\alpha|=m} \int_\Omega |\partial^\alpha u(x)|^p dx\Big)^{1/p} \quad ,$$

for $p < \infty$, and we make the above modification when $p = \infty$.

When $p = 2$, $W^{m,2}(\Omega)$ is usually denoted by $H^m(\Omega)$, and if there is no ambiguity, we drop the subscript $p = 2$ when refering to its norm and seminorm. $H^m(\Omega)$ is a Hilbert space for the scalar product :

$$(1.3) \qquad (u,v)_{m,\Omega} = \sum_{|\alpha|\leq m} \int_\Omega \partial^\alpha u(x) \partial^\alpha v(x) dx .$$

In particular, we write the scalar product of $L^2(\Omega)$ with no subscript at all.

As $\mathcal{D}(\Omega) \subset H^m(\Omega)$, we define

$$(1.4) \qquad H_0^m(\Omega) = \overline{\mathcal{D}(\Omega)}^{H^m(\Omega)} ,$$

i.e. $H_0^m(\Omega)$ is the closure of $\mathcal{D}(\Omega)$ for the norm $\| \cdot \|_{m,\Omega}$. We denote by $H^{-m}(\Omega)$ the dual space of $H_0^m(\Omega)$ normed by :

$$(1.5) \qquad \|f\|_{-m,\Omega} = \sup_{\substack{v \in H_0^m(\Omega) \\ v \neq 0}} \frac{|< f,v >|}{\|v\|_{m,\Omega}} .$$

The following lemma characterizes the functionals of $H^{-m}(\Omega)$.

LEMMA 1.1.

A distribution f belongs to $H^{-m}(\Omega)$ if and only if there exist functions f_α in $L^2(\Omega)$, for $|\alpha| \leq m$, such that

$$f = \sum_{|\alpha| \leq m} \partial^\alpha f_\alpha .$$

THEOREM 1.1. (Poincaré-Friedrichs' inequality)

If Ω is connected and *bounded at least in one direction*, then for each $m \in \mathbb{N}$, there exists a constant C_m such that

$$\|v\|_{m,\Omega} \leq C_m |v|_{m,\Omega} \qquad \forall v \in H^m_o(\Omega) . \tag{1.6}$$

Hence the mapping $v \longmapsto |v|_{m,\Omega}$ is a norm on $H^m_o(\Omega)$ equivalent to $\|v\|_{m,\Omega}$.

In order to study more closely the boundary values of functions of $H^m(\Omega)$, we assume that Γ , the boundary of Ω , is *bounded and Lipschitz continuous* - i.e. Γ can be represented parametrically by Lipschitz continuous functions. Let $d\sigma$ denote the surface measure on Γ and let $L^2(\Gamma)$ be the space of square integrable functions on Γ with respect to $d\sigma$, equipped with the norm

$$\|v\|_{o,\Gamma} = \{\int_\Gamma (v(\sigma))^2 d\sigma\}^{1/2} .$$

THEOREM 1.2.

1°) $\mathcal{D}(\overline{\Omega})$ is dense in $H^1(\Omega)$.

2°) There exists a constant C such that

$$\|\gamma_o\varphi\|_{o,\Gamma} \leq C\|\varphi\|_{1,\Omega} \qquad \forall \varphi \in \mathcal{D}(\overline{\Omega}) , \tag{1.7}$$

where $\gamma_o\varphi$ denotes the value of φ on Γ .

It follows from Theorem 1.2. that the mapping γ_o defined on $\mathcal{D}(\overline{\Omega})$ can be extended by continuity to a mapping, still called γ_o , from $H^1(\Omega)$ into $L^2(\Gamma)$, i.e. $\gamma_o \in \mathcal{L}(H^1(\Omega) ; L^2(\Gamma))$. By extension, $\gamma_o\varphi$ is called the boundary value of φ on Γ ; to simplify notations, we drop the prefix γ_o when it is clearly implied.

THEOREM 1.3.

1°) $\mathrm{Ker}(\gamma_o) = H_o^1(\Omega)$.

2°) The range space of γ_o is a proper and dense subspace of $L^2(\Gamma)$, called $H^{1/2}(\Gamma)$.

For μ in $H^{1/2}(\Gamma)$, we define

$$\|\mu\|_{1/2,\Gamma} = \inf_{\substack{v \in H^1(\Omega) \\ \gamma_o v = \mu}} \|v\|_{1,\Omega} . \tag{1.8}$$

The mapping $\mu \longmapsto \|\mu\|_{1/2,\Gamma}$ is a norm on $H^{1/2}(\Gamma)$, and $H^{1/2}(\Gamma)$ is a Hilbert space for this norm. Let $H^{-1/2}(\Gamma)$ be the corresponding dual space of $H^{1/2}(\Gamma)$, normed by

$$\|\mu^*\|_{-1/2,\Gamma} = \sup_{\substack{\mu \in H^{1/2}(\Gamma) \\ \mu \neq 0}} \frac{|\langle \mu^*,\mu \rangle|}{\|\mu\|_{1/2,\Gamma}} , \tag{1.9}$$

where again $\langle .,. \rangle$ denotes the duality between $H^{-1/2}(\Gamma)$ and $H^{1/2}(\Gamma)$. We remark that $\langle .,. \rangle$ is an extension of the scalar product of $L^2(\Gamma)$ in the sense that when $\mu^* \in L^2(\Gamma)$, we can identify $\langle \mu^*,\mu \rangle$ with $\int_\Gamma \mu^*(\sigma)\mu(\sigma)\, d\sigma$.

Let $\vec{\nu} = (\nu_1 ,\dots, \nu_n)$ be the unit outward normal to Γ which exists almost everywhere on Γ thanks to the hypothesis of Lipschitz continuity. If v is a function in $H^2(\Omega)$, we define its normal derivative by :

$$\frac{\partial v}{\partial \nu} = \sum_{i=1}^{n} \nu_i \gamma_o \left(\frac{\partial v}{\partial x_i}\right) . \tag{1.10}$$

It can be proved that the mapping $v \longmapsto \frac{\partial v}{\partial \nu} \in \mathcal{L}(H^2(\Omega) ; H^{1/2}(\Gamma))$. Moreover, we can characterize $H_o^2(\Omega)$ as follows :

THEOREM 1.4.

$$H_o^2(\Omega) = \{v \in H^2(\Omega) ; \gamma_o v = 0 \ \underline{\text{and}}\ \frac{\partial v}{\partial \nu} = 0\} .$$

When Γ is sufficiently smooth, the range space of γ_o can also be extended as follows. For $m \in \mathbb{N}$, $m \geqslant 1$, we define $H^{m-1/2}(\Gamma)$ as the image of $H^m(\Omega)$ by the transformation γ_o, equipped with the norm :

$$\|f\|_{m-1/2,\Gamma} = \inf_{\substack{v \in H^m(\Omega) \\ \gamma_o v = f}} \|v\|_{m,\Omega} .$$

Then, it can be checked that $\frac{\partial u}{\partial \nu} \in H^{m-3/2}(\Gamma)$ for u in $H^m(\Omega)$, and the following result holds :

<u>THEOREM 1.5.</u>

<u>The mapping</u> $u \longmapsto \{\gamma_o u, \frac{\partial u}{\partial \nu}\}$ <u>defined on</u> $H^m(\Omega)$ <u>is *onto*</u> $H^{m-1/2}(\Gamma) \times H^{m-3/2}(\Gamma)$.

We close this section with two useful applications of the Green's formula.

<u>LEMMA 1.2.</u>

1°) <u>Let</u> u <u>and</u> $v \in H^1(\Omega)$. <u>Then, for</u> $1 \leqslant i \leqslant n$,

$$\int_\Omega u \frac{\partial v}{\partial x_i}\, dx = - \int_\Omega \frac{\partial u}{\partial x_i} v\, dx + \int_\Gamma u v \nu_i\, d\sigma . \tag{1.11}$$

2°) <u>Moreover, if</u> $u \in H^2(\Omega)$, <u>then</u>

$$\sum_{i=1}^n \int_\Omega \frac{\partial u}{\partial x_i} \frac{\partial v}{\partial x_i}\, dx = - \sum_{i=1}^n \int_\Omega \frac{\partial^2 u}{\partial x_i^2} v\, dx + \sum_{i=1}^n \int_\Gamma \nu_i \frac{\partial u}{\partial x_i} v\, d\sigma . \tag{1.12}$$

Adopting the usual notations :

$$\Delta u = \sum_{i=1}^n \frac{\partial^2 u}{\partial x_i^2} , \quad \overrightarrow{\text{grad}}\, u = (\frac{\partial u}{\partial x_1}, \ldots, \frac{\partial u}{\partial x_n}) ,$$

(1.12) becomes :

$$(\overrightarrow{\text{grad}}\, u, \overrightarrow{\text{grad}}\, v) = - (\Delta u, v) + \int_\Gamma \frac{\partial u}{\partial \nu} v\, d\sigma . \tag{1.13}$$

1.2. <u>Abstract elliptic theory</u>

This section gives a brief account of a fundamental tool used in studying linear partial differential equations of elliptic type.

Let V be a real Hilbert space with norm denoted by $\|\cdot\|_V$; let V' be its dual space and let $<.,.>$ denote the duality between V' and V .
Let $(u,v) \longmapsto a(u,v)$ be a real bilinear form on $V \times V$, ℓ an element of V' and consider the following problem :

$$\text{(P)} \quad \begin{cases} \text{Find } u \in V \text{ such that} \\ a(u,v) = < \ell, v > \quad \forall\, v \in V . \end{cases}$$

The following theorem is due to Lax and Milgram [35] .

THEOREM 1.6.

We assume that a is continuous and elliptic on V, i.e. there exist two constants M and $\alpha > 0$ such that

$$|a(u,v)| \leq M \|u\|_V \|v\|_V \qquad \forall\, u,v \in V \tag{1.14}$$

and

$$a(v,v) \geq \alpha \|v\|_V^2 \qquad \forall\, v \in V . \tag{1.15}$$

Then problem (P) has one and only one solution u in V. Moreover, the mapping $\ell \longmapsto u$ is an *isomorphism* from V' *onto* V .

COROLLARY 1.1.

When a is symmetric - i.e. $a(u,v) = a(v,u)$ $\forall\, u, v \in V$ - then the solution u of (P) is also the only element of V that minimizes the following quadratic functional (also called energy functional) on V :

$$J(v) = \frac{1}{2} a(v,v) - < \ell,v > . \tag{1.16}$$

1.3. Example 1 : Dirichlet's harmonic problem

In all the examples, we assume that Ω is bounded and Γ Lipschitz continuous.

Consider the following non-homogeneous Dirichlet's problem :

(D) Given f in $H^{-1}(\Omega)$ and g in $H^{1/2}(\Gamma)$, find a function u that satisfies

$$- \Delta u = f \text{ in } \Omega \tag{1.17}$$

$$u = g \text{ on } \Gamma . \tag{1.18}$$

Let us formulate this problem in terms of problem (P). We set $V = H_0^1(\Omega)$ and

$$a(u,v) = (\overrightarrow{\text{grad}}\, u, \overrightarrow{\text{grad}}\, v).$$

It is clear that a is continuous on $[H_0^1(\Omega)]^2$, and owing to Theorem 1.1,

$$a(v,v) = \|\overrightarrow{\text{grad}}\, v\|_{0,\Omega}^2 = |v|_{1,\Omega}^2 \geq C_1 \|v\|_{1,\Omega}^2 .$$

Besides that since $H^{1/2}(\Gamma)$ is the range space of γ_0, let u_0 in $H^1(\Omega)$ satisfy $\gamma_0 u_0 = g$, and examine the following problem :

$$(D')\quad \begin{cases} \text{Find } u \text{ in } H^1(\Omega) \text{ such that} \\ (1.19) \quad u - u_o \in H^1_o(\Omega) \\ (1.20) \quad a(u-u_o,v) = < f,v > - a(u_o,v) \quad \forall\, v \in H^1_o(\Omega)\ . \end{cases}$$

Since a is continuous, the mapping $v \overset{\ell}{\longmapsto} < f,v > - a(u_o,v)$ belongs to $H^{-1}(\Omega)$. Therefore, thanks to the Lax-Milgram theorem, problem (D') has one and only one solution u in $H^1(\Omega)$.

It remains only to prove that u may be characterized as the unique solution of problem (D). Taking $v \in \mathcal{D}(\Omega)$ in (1.20) gives :

$$a(u,v) = - < \Delta u,v > = < f,v > \qquad \forall\, v \in \mathcal{D}(\Omega)\ .$$

Hence u satisfies

$$(D_1)\quad \begin{cases} (1.19) \quad u-u_o \in H^1_o(\Omega)\ , \\ (1.17) \quad -\Delta u = f \text{ in } H^{-1}(\Omega)\ . \end{cases}$$

Conversely, every solution of (D_1) is a solution of (D') by the density of $\mathcal{D}(\Omega)$ in $H^1_o(\Omega)$. But

$$u - u_o \in H^1_o(\Omega) \quad \text{iff} \quad \gamma_o u = g\ ,$$

therefore problems (D_1) and (D) are the same.

As far as the regularity of u is concerned, we know, from the Lax-Milgram's theorem that the mapping $\ell \longmapsto u - u_o$ is an isomorphism from $H^{-1}(\Omega)$ onto $H^1(\Omega)$. Therefore,

$$\| u-u_o \|_{1,\Omega} \leq C_2 \| \ell \|_{-1,\Omega}\ .$$

Clearly,

$$\| \ell \|_{-1,\Omega} \leq \| f \|_{-1,\Omega} + \| u_o \|_{1,\Omega}\ .$$

Hence

$$\| u \|_{1,\Omega} \leq C_3 \{ \| f \|_{-1,\Omega} + \| u_o \|_{1,\Omega} \}$$

$\forall\, u_o \in H^1(\Omega)$ such that $\gamma_o u_o = g$. From definition (1.8) this implies that

$$\| u \|_{1,\Omega} \leq C_3 \{ \| f \|_{-1,\Omega} + \| g \|_{1/2,\Gamma} \}\ .$$

Thus, we have proved the following proposition :

PROPOSITION 1.1.

Problem (D) has one and only one solution u in $H^1(\Omega)$ and

(1.21) $$\|u\|_{1,\Omega} \leq C\{\|f\|_{-1,\Omega} + \|g\|_{1/2,\Gamma}\} ,$$

i.e. u depends continuously upon the data of (D).

Remarks 1.1.

1°) Let $m \in \mathbb{N}$, $m \geq 1$. When Γ is sufficiently smooth, it can be shown that if $f \in H^{m-2}(\Omega)$ and $g \in H^{m-1/2}(\Gamma)$, then $u \in H^m(\Omega)$ and

(1.22) $$\|u\|_{m,\Omega} \leq C(\|f\|_{m-2,\Omega} + \|g\|_{m-1/2,\Gamma}) .$$

2°) When Γ is only Lipschitz continuous, the same result is still valid for $m = 2$, provided Ω is convex. ■

1.4. Example 2 : Neumann's harmonic problem

Here, we assume in addition that Ω is connected and we deal with the non-homogeneous Neumann's problem :

(N) Find u such that :

(1.23) $$-\Delta u = f \quad \text{in } \Omega ,$$

(1.24) $$\frac{\partial u}{\partial \nu} = g \quad \text{on } \Gamma ,$$

where $f \in L^2(\Omega)$ and $g \in H^{-1/2}(\Gamma)$ satisfy the relation :

(1.25) $$\int_\Omega f\,dx + \langle g, 1 \rangle_\Gamma = 0 .$$

Since problem (N) only involves the derivatives of u, it is clear that its solution is never unique. We turn the difficulty by seeking u in the quotient space $H^1(\Omega)/\mathbb{R}$ equipped with the quotient norm :

(1.26) $$\|\dot{v}\|_{H^1(\Omega)/\mathbb{R}} = \inf_{v \in \dot{v}} \|v\|_{1,\Omega} .$$

The theorem below states an important property of this space ; its proof can be found in Nečas [39] .

THEOREM 1.7.

The space $H^1(\Omega)/\mathbb{R}$ is a Hilbert space for the quotient norm (1.26). Moreover, on this space the seminorm $\dot{v} \longmapsto |v|_{1,\Omega}$ is a norm, equivalent to (1.26).

With this space, we can put problem (N) in the abstract setting of problem (P). Let $V = H^1(\Omega)/\mathbb{R}$,

$$a(\dot{u},\dot{v}) = (\overrightarrow{\text{grad}}\ u, \overrightarrow{\text{grad}}\ v) \qquad \forall\, u \in \dot{u}\ ,\ v \in \dot{v}$$

and

$$\ell : \dot{v} \longmapsto \int_\Omega fv\,dx + \langle g,v \rangle_\Gamma \qquad \forall\, v \in \dot{v}\,. \tag{1.27}$$

Note that the right-hand side of (1.27) is independent of the particular $v \in \dot{v}$ thanks to the compatibility condition (1.25). Furthermore, $\ell \in V'$ because, owing to (1.8), we have :

$$\left|\int_\Omega fv\,dx + \langle g,v \rangle_\Gamma\right| \leq (\|f\|_{o,\Omega} + \|g\|_{-1/2,\Gamma}) \inf_{v \in \dot{v}} \|v\|_{1,\Omega}\,.$$

Thus

$$\|\ell\|_{V'} \leq \|f\|_{o,\Omega} + \|g\|_{-1/2,\Gamma}\,. \tag{1.28}$$

Obviously, $a(\dot{u},\dot{v})$ is continuous on $V \times V$, and by virtue of Theorem 1.7,

$$a(\dot{v},\dot{v}) = |v|^2_{1,\Omega} \geq C_1 \|\dot{v}\|^2_{H^1(\Omega)/\mathbb{R}}\,.$$

Hence, by the Lax-Milgram's theorem, the following problem

$$(\text{N}')\ \begin{cases} \text{Find } \dot{u} \text{ in } H^1(\Omega)/\mathbb{R} \text{ satisfying} \\ (1.29) \qquad a(\dot{u},\dot{v}) = \langle \ell,\dot{v} \rangle \qquad \forall\, \dot{v} \in H^1(\Omega)/\mathbb{R}\ , \end{cases}$$

has a unique solution $\dot{u} \in H^1(\Omega)/\mathbb{R}$.

Let us interpret problem (N'). When v is restricted to $\mathcal{D}(\Omega)$, (1.29) yields :

$$-\Delta u = f \quad \text{in} \quad L^2(\Omega) \qquad \forall\, u \in \dot{u}\,. \tag{1.30}$$

Next, by taking the scalar product of (1.30) with v and comparing with (1.29), we find :

$$(\overrightarrow{\text{grad}}\ u, \overrightarrow{\text{grad}}\ v) = (-\Delta u, v) + \langle g,v \rangle_\Gamma \qquad \forall\, v \in H^1(\Omega)\,. \tag{1.31}$$

Therefore, problem (N') is equivalent to :

find u in $H^1(\Omega)$ satisfying (1.30) and (1.31).

It remains to interpret (1.31) as a boundary condition. At the present stage this cannot be done without assuming that $u \in H^2(\Omega)$. Then Green's formula (1.13) yields :

$$\int_\Gamma \frac{\partial u}{\partial \nu} v \, d\sigma = \langle g,v \rangle_\Gamma \qquad \forall v \in H^1(\Omega) ,$$

i.e.

$$\frac{\partial u}{\partial \nu} = g \text{ on } \Gamma .$$

As u is supposed to belong to $H^2(\Omega)$, this implies in particular that $g \in H^{1/2}(\Gamma)$. In that case, problems (N) and (N') are equivalent. Of course this is not entirely satisfactory in that the existence of a solution of problem (N) is subjected to the regularity of the solution of (N'). Further on, with more powerful tools, we shall be able to eliminate this regularity hypothesis.

Now, let us examine the regularity of u . According to the Lax-Milgram' s theorem, (1.28) and the equivalence Theorem 1.7, we obtain :

$$|u|_{1,\Omega} \leqslant C_2(\|f\|_{o,\Omega} + \|g\|_{-1/2,\Gamma}) .$$

We have thus proved the following result.

PROPOSITION 1.2.

Let the solution $\dot{u}$ of problem (N') belong to $H^2(\Omega)/\mathbb{R}$. Then $\dot{u}$ is the only solution of problem (N) and each $u \in \dot{u}$ is continuous with respect to the data, i.e.

$$|u|_{1,\Omega} \leqslant C(\|f\|_{o,\Omega} + \|g\|_{-1/2,\Gamma}) . \tag{1.32}$$

Remark 1.2.

As in the previous example, if Γ is very smooth and if $f \in H^{m-2}(\Omega)$ and $g \in H^{m-3/2}(\Gamma)$ with $m \geqslant 2$, then it can be shown that $\dot{u} \in H^m(\Omega)/\mathbb{R}$ and

$$|u|_{m,\Omega} \leqslant C(\|f\|_{m-2,\Omega} + \|g\|_{m-3/2,\Gamma}) \text{ for every } u \in \dot{u} . \tag{1.33}$$

■

1.5. Example 3 : Dirichlet's biharmonic problem

Consider the non-homogeneous fourth order problem :

For f given in $H^{-2}(\Omega)$, g_1 given in $H^{3/2}(\Gamma)$ and g_2 in $H^{1/2}(\Gamma)$,

(B) find u such that :

$$\Delta^2 u = f \text{ in } \Omega, \tag{1.34}$$

$$u = g_1 \text{ on } \Gamma \tag{1.35}$$

and

$$\frac{\partial u}{\partial \nu} = g_2 \text{ on } \Gamma. \tag{1.36}$$

The function space naturally attached to this problem is $H_o^2(\Omega)$ and the bilinear form is :

$$a(u,v) = (\Delta u, \Delta v).$$

This form is elliptic on $H_o^2(\Omega)$ because the mapping $v \longmapsto \|\Delta v\|_{o,\Omega}$ is a norm on $H_o^2(\Omega)$ equivalent to the norm $\|\cdot\|_{2,\Omega}$. Indeed, for v in $\mathcal{D}(\Omega)$, we can easily show by integrating by parts and interchanging derivatives that

$$\|\Delta v\|_{o,\Omega}^2 = |v|_{2,\Omega}^2. \tag{1.37}$$

By density, the same result holds for the functions of $H_o^2(\Omega)$. The equivalence follows from Theorem 1.1.

According to Theorem 1.5, if Γ is smooth enough, there exists a function u_o in $H^2(\Omega)$ such that

$$\gamma_o u_o = g_1 \quad , \quad \frac{\partial u_o}{\partial \nu} = g_2 \text{ on } \Gamma. \tag{1.38}$$

Thus we turn to the following problem :

(B') Find u in $H^2(\Omega)$ such that

$$u - u_o \in H_o^2(\Omega), \tag{1.39}$$

$$a(u-u_o,v) = \langle f,v \rangle - a(u_o,v) \quad \forall v \in H_o^2(\Omega). \tag{1.40}$$

By the Lax-Milgram's theorem, problem (B') has exactly one solution u in $H^2(\Omega)$. Owing to (1.38) and (1.39), u satisfies the boundary conditions

$$\gamma_o u = g_1 \quad , \quad \frac{\partial u}{\partial \nu} = g_2 \text{ on } \Gamma.$$

Besides that, by restricting the test functions of (1.40) to $\mathcal{D}(\Omega)$, we find

$$\Delta^2 u = f \quad \text{in} \quad H^{-2}(\Omega).$$

Therefore, u is a solution of (B).

Conversely, as in the case of the harmonic operator, we can show that problem (B) has at most one solution in $H^2(\Omega)$.

From (1.40) and the equivalence of norms, we derive the bound

$$\| u \|_{2,\Omega} \leq C_1(\| f \|_{-2,\Omega} + \| u_o \|_{2,\Omega}) \quad \forall\, u_o \text{ satisfying (1.38)},$$

i.e.
$$\| u \|_{2,\Omega} \leq C_2(\| f \|_{-2,\Omega} + \| g_1 \|_{3/2,\Gamma} + \| g_2 \|_{1/2,\Gamma}).$$

These results are summed up in the proposition below :

PROPOSITION 1.3.

If Γ is sufficiently smooth, problem (B) has exactly one solution u in $H^2(\Omega)$, bounded as follows :

$$\| u \|_{2,\Omega} \leq C(\| f \|_{-2,\Omega} + \| g_1 \|_{3/2,\Gamma} + \| g_2 \|_{1/2,\Gamma}). \tag{1.41}$$

Remark 1.3.

When Γ is sufficiently differentiable it can be shown that, if $f \in H^{m-4}(\Omega)$, $g_1 \in H^{m-1/2}(\Gamma)$ and $g_2 \in H^{m-3/2}(\Gamma)$ for $m \geq 2 \in \mathbb{N}$, then $u \in H^m(\Omega)$ and

$$\| u \|_{m,\Omega} \leq C\,(\| f \|_{m-4,\Omega} + \| g_1 \|_{m-1/2,\Gamma} + \| g_2 \|_{m-3/2,\Gamma}). \quad \blacksquare$$

§ 2 - SOME FUNCTION SPACES

Throughout this paragraph, we assume that Ω is an open subset of $\mathbb{R}^n$ with a bounded and Lipschitz continuous boundary Γ. We introduce here special Hilbert spaces that are particularly well suited to incompressible flows and other problems arising in mechanics. Several results are stated without proof ; these can be found in the book by Duvaut and Lions [22]. In addition, the reader will find in the Appendix an alternate shorter version of the proofs of Theorems 2.1 and 2.3.

2.1. The space H(div ; Ω)

From now on, we shall often deal with vector-valued functions. We shall distinguish vectors by means of arrows and extend naturally all the previous norms to vectors as follows : if $\vec{v} = (v_1, \ldots, v_n)$ then

$$\|\vec{v}\|_{m,p,\Omega} = \left(\sum_{i=1}^{n} \|v_i\|^p_{m,p,\Omega}\right)^{1/p}.$$

For such vectors, we define the divergence operator by

$$\operatorname{div} \vec{v} = \sum_{i=1}^{n} \frac{\partial v_i}{\partial x_i}.$$

Then, we introduce the following spaces :

$$H(\operatorname{div} ; \Omega) = \{\vec{v} \in (L^2(\Omega))^n ; \operatorname{div} \vec{v} \in L^2(\Omega)\},$$

normed by :

$$\|\vec{v}\|_{H(\operatorname{div} ; \Omega)} = \{\|\vec{v}\|^2_{o,\Omega} + \|\operatorname{div} \vec{v}\|^2_{o,\Omega}\}^{1/2} ; \tag{2.1}$$

and

$$H_o(\operatorname{div} ; \Omega) = \overline{(\mathcal{D}(\Omega))^n}^{H(\operatorname{div} ; \Omega)}.$$

Clearly, H(div ; Ω) is a Hilbert space for the norm (2.1).

THEOREM 2.1.

$$[\mathcal{D}(\overline{\Omega})]^n \text{ is dense in } H(\operatorname{div} ; \Omega).$$

PROOF.

Let $\vec{u} \in H(\operatorname{div} ; \Omega)$. First, let us show that there exists a sequence of functions of H(div ; Ω) with compact support, that tends to $\vec{u}$.

Let a denote any positive real number and let φ_a be a function of $\mathcal{D}(\mathbb{R}^n)$ such that :

$$\varphi_a(x) = \begin{cases} 1 & \text{for } |x| \leqslant a \\ 0 & \text{for } |x| \geqslant 2a \end{cases}$$

and $0 \leqslant \varphi_a \leqslant 1$ everywhere in $\mathbb{R}^n$. As φ_a is smooth, $\varphi_a \vec{u} \in H(\operatorname{div} ; \Omega)$; moreover $\lim_{a \to \infty} \varphi_a \vec{u} = \vec{u}$ in H(div ; Ω) and the support of $\varphi_a \vec{u}$ is compact. Hence $\varphi_a \vec{u}$ is the desired sequence.

From the above, we can assume that $\overline{\Omega}$ is compact. We wish to extend $\vec{u}$

to $\mathbb{R}^n$ so that the extended function belongs to $H(\text{div} ; \mathbb{R}^n)$. Then we can apply a classical procedure of regularization to show that $\mathcal{D}(\mathbb{R}^n)$ is dense in $H(\text{div} ; \mathbb{R}^n)$.

1) Since Ω is eventually multiply-connected, we shall denote by Γ_o the exterior boundary of Ω (cf. figure 1) and by Γ_i , $1 \leq i \leq p$, the other components of Γ . Then let $\mathcal{O}$ be a simply-connected, bounded open set with a smooth boundary $\partial\mathcal{O}$, such that $\bar{\Omega} \subset \mathcal{O}$. Thus $\mathcal{O}-\bar{\Omega}$ is an open set, not necessarily connected, bounded by Γ and $\partial\mathcal{O}$. Let Ω_i , $1 \leq i \leq p$ (resp. Ω_o) denote the component of $\mathcal{O}-\bar{\Omega}$ bounded by Γ_i (resp. Γ_o and $\partial\mathcal{O}$). Let

$$V = \{\text{classes } \dot{v} \, ; \dot{v}|_{\Omega_i} \in H^1(\Omega_i)/\mathbb{R}\}$$

and consider the following problem : Find $\dot{w} \in V$ satisfying

$$(N) \begin{cases} \displaystyle\int_{\mathcal{O}-\bar{\Omega}} \overrightarrow{\text{grad}}\, \dot{w} \cdot \overrightarrow{\text{grad}}\, \dot{v}\, dx = \langle \ell, \dot{v} \rangle \quad \forall \dot{v} \in V , \\ \text{where} \\ (2.2) \quad \langle \ell, \dot{v} \rangle = \displaystyle\sum_{i=0}^{p} \frac{1}{\text{meas}(\Omega_i)} \left(\int_\Omega \overrightarrow{\text{grad}}\, \pi e_i \cdot \vec{u}\, dx + \int_\Omega \pi e_i \,\text{div}\, \vec{u}\, dx \right) \int_{\Omega_i} v\, dx \\ \qquad - \displaystyle\int_\Omega \overrightarrow{\text{grad}}\, \pi v . \vec{u}\, dx - \int_\Omega \pi v \,\text{div}\, \vec{u}\, dx , \end{cases}$$

where v is any representative of $\dot{v}$, $e_i|_{\Omega_j} = \delta_i^j$ for $0 \leq i,j \leq p$ and $\pi\varphi$ is any extension of φ in $H^1(\mathcal{O})$ that coincides with φ in $\mathcal{O}-\bar{\Omega}$.

We observe that problem (N) is a generalization of the non-homogeneous Neumann's problem studied in § 1.4. Indeed a straightforward extension of Theorem 1.7 shows that the seminorm

$$|\dot{v}| = |v|_{1,\mathcal{O}-\bar{\Omega}}$$

is a norm on V, equivalent to

$$\|\dot{v}\| = \left(\sum_{i=0}^{p} \|\dot{v}\|^2_{H^1(\Omega_i)/\mathbb{R}}\right)^{1/2}.$$

Moreover, the right-hand side of (2.2) is independent of π since

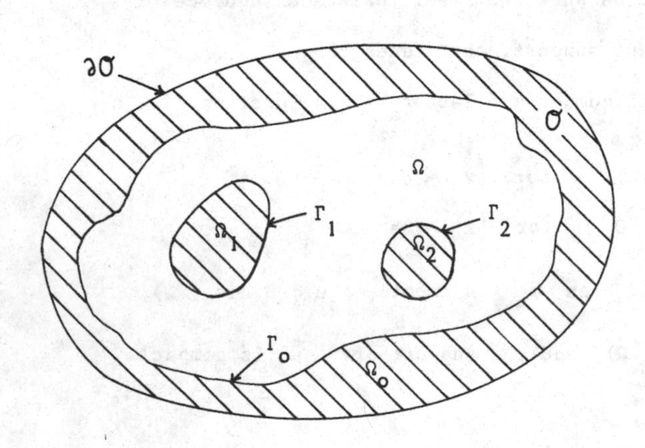

FIGURE 1

$$\int_\Omega \overrightarrow{\text{grad}}\,\varphi.\vec{u}\,dx - \int_\Omega \varphi \,\text{div}\,\vec{u}\,dx = 0 \quad \forall \varphi \in H_0^1(\Omega) .$$

Furthermore, if $\dot{v} = \dot{0}$ then $v = \sum_{i=0}^{p} c_i e_i$, where the c_i are constants and clearly $< \ell,\dot{0} > = 0$. Finally,

$$| < \ell,\dot{v} >| \leqslant C_1 \|\vec{u}\|_{H(\text{div}\,;\,\Omega)} \|v\|_{1,\mathcal{O}-\bar{\Omega}} \qquad \forall\, v \in \dot{v} ,$$

since we can assume that π is continuous. Hence $\ell \in V'$. As a consequence, problem (N) has a unique solution $\dot{w} \in V$ that satisfies :

$$\Delta w_{|\Omega_i} = \frac{-1}{\text{meas}(\Omega_i)}\left(\int_\Omega \overrightarrow{\text{grad}}\,\pi e_i.\vec{u}\,dx + \int_\Omega \pi e_i \text{div}\,\vec{u}\,dx\right) , \quad 0 \leqslant i \leqslant p.$$

Now, as a distribution, $\text{div}\,(\overrightarrow{\text{grad}}\,w) = \Delta w$. Therefore $\overrightarrow{\text{grad}}\,w \in H(\text{div}\,;\mathcal{O}-\bar{\Omega})$.

Then, let us extend $\vec{u}$ as follows :

$$\tilde{\vec{u}} = \begin{cases} \vec{u} \text{ in } \bar{\Omega} \\ \overrightarrow{\text{grad}}\,w \text{ in } \mathcal{O}-\bar{\Omega} \\ 0 \text{ elsewhere.} \end{cases}$$

Obviously, $\tilde{\vec{u}} \in (L^2(\mathbb{R}^n))^n$, and we must check that $\text{div}\,\tilde{\vec{u}} \in L^2(\mathbb{R}^n)$. As a distribution $\text{div}\,\tilde{\vec{u}}$ is defined by

$$< \text{div}\,\tilde{\vec{u}},\varphi > = - \int_{\mathbb{R}^n} \tilde{\vec{u}}\cdot\overrightarrow{\text{grad}}\,\varphi\,dx \qquad \forall \varphi \in \mathcal{D}(\mathbb{R}^n) .$$

But

$$\int_{\mathbb{R}^n} \tilde{\vec{u}}\cdot\overrightarrow{\text{grad}}\,\varphi\,dx = \int_\Omega \vec{u}\cdot\overrightarrow{\text{grad}}\,\varphi\,dx + \int_{\mathcal{O}-\bar{\Omega}} \overrightarrow{\text{grad}}\,w\cdot\overrightarrow{\text{grad}}\,\varphi\,dx .$$

Therefore, by (2.2)

$$(2.3) \qquad \int_{\mathbb{R}^n} \tilde{\vec{u}}\cdot\overrightarrow{\text{grad}}\,\varphi\,dx = \int_\Omega \vec{u}\cdot\overrightarrow{\text{grad}}\,\varphi\,dx + \sum_{i=0}^{p} \frac{1}{\text{meas}(\Omega_i)}\left(\int_\Omega \overrightarrow{\text{grad}}\,\pi e_i\cdot\vec{u}\,dx + \int_\Omega \pi\, e_i\,\text{div}\,\vec{u}\,dx\right) \int_{\Omega_i} \varphi\,dx - \int_\Omega \overrightarrow{\text{grad}}\,\varphi\cdot\vec{u}\,dx - \int_\Omega \varphi\,\text{div}\,\vec{u}\,dx.$$

As the right-hand side of (2.3) is bounded by $C_2 \|\vec{u}\|_{H(\text{div}\,;\,\Omega)} \|\varphi\|_{o,\mathcal{O}}$, we infer that $\text{div}\,\tilde{\vec{u}} \in L^2(\mathbb{R}^n)$.

2) Let δ_ε be a regularizing sequence of $\mathcal{D}(\mathbb{R}^n)$, i.e. $\delta_\varepsilon(x) \geqslant 0$,

$\int_{\mathbb{R}^n} \delta_\varepsilon(x)\,dx = 1$ and $\lim_{\varepsilon \to 0} \delta_\varepsilon = \delta$. Consider the sequence $\delta_\varepsilon \star \tilde{\vec{u}}$.
The properties of the convolution imply that $\delta_\varepsilon \star \tilde{\vec{u}} \in (\mathcal{D}(\mathbb{R}^n))^n$. By taking limits we find that :

$$\lim_{\varepsilon \to 0} \delta_\varepsilon \star \tilde{\vec{u}} = \delta \star \tilde{\vec{u}} = \tilde{\vec{u}} \quad \text{in } [L^2(\mathbb{R}^n)]^n$$

and

$$\lim_{\varepsilon \to 0} \operatorname{div}(\delta_\varepsilon \star \tilde{\vec{u}}) = \lim_{\varepsilon \to 0} (\delta_\varepsilon \star \operatorname{div} \tilde{\vec{u}}) = \operatorname{div} \tilde{\vec{u}} \quad \text{in } L^2(\mathbb{R}^n) .$$

Therefore the restriction to $\bar{\Omega}$ of $\delta_\varepsilon \star \tilde{\vec{u}}$ is a function of $(\mathcal{D}(\bar{\Omega}))^n$ that converges toward $\vec{u}$ for the norm of $H(\operatorname{div} ; \Omega)$. ■

Remark 2.1.

We shall see later on that, in this theorem, $\vec{u}$ is extended by matching its normal component with that of its extension. This process will be used several times. ■

Again, let $\vec{\nu}$ denote the unit exterior normal along Γ . The next theorem concerns the boundary values of functions of $H(\operatorname{div} ; \Omega)$.

THEOREM 2.2.

The mapping $\gamma_\nu : \vec{v} \longmapsto \vec{v}\cdot\vec{\nu}_{|\Gamma}$ defined on $(\mathcal{D}(\bar{\Omega}))^n$ can be extended by continuity to a linear and continuous mapping, still denoted by γ_ν , from $H(\operatorname{div} ; \Omega)$ into $H^{-1/2}(\Gamma)$.

PROOF.

Let $\varphi \in \mathcal{D}(\bar{\Omega})$ and $\vec{v} \in (\mathcal{D}(\bar{\Omega}))^n$. The following Green's formula holds :

$$(2.4) \qquad (\vec{v}, \overrightarrow{\operatorname{grad}}\,\varphi) + (\operatorname{div} \vec{v}, \varphi) = \int_\Gamma \varphi \vec{v}\cdot\vec{\nu}\, d\sigma .$$

As $\mathcal{D}(\bar{\Omega})$ is dense in $H^1(\Omega)$, (2.4) is still valid for φ in $H^1(\Omega)$ and $\vec{v}$ in $(\mathcal{D}(\bar{\Omega}))^n$. Therefore

$$(2.5) \qquad \left|\int_\Gamma \varphi\vec{v}\cdot\vec{\nu}\, d\sigma\right| \leq \|\vec{v}\|_{H(\operatorname{div} ; \Omega)} \|\varphi\|_{1,\Omega} \quad \forall\, \varphi \in H^1(\Omega) , \ \forall\, \vec{v} \in (\mathcal{D}(\bar{\Omega}))^n .$$

Now, let μ be any element of $H^{1/2}(\Gamma)$. Then there exists an element φ of

$H^1(\Omega)$ such that $\gamma_o\varphi = \mu$. Hence (2.5) implies that

$$\left|\int_\Gamma \mu\, \vec{v}\cdot\vec{\nu}\, d\sigma\right| \leq \|\vec{v}\|_{H(\mathrm{div}\,;\,\Omega)} \|\mu\|_{1/2,\Gamma} \quad \forall\, \mu \in H^{1/2}(\Gamma),\ \forall\, \vec{v} \in (\mathcal{D}(\overline{\Omega}))^n .$$

Thus
$$\|\vec{v}\cdot\vec{\nu}\|_{-1/2,\Gamma} \leq \|\vec{v}\|_{H(\mathrm{div}\,;\,\Omega)} .$$

Hence, the linear mapping $\gamma_\nu : \vec{v} \longmapsto \vec{v}\cdot\vec{\nu}_{|\Gamma}$ defined on $(\mathcal{D}(\overline{\Omega}))^n$ is continuous for the norm of $H(\mathrm{div}\,;\,\Omega)$. Since $(\mathcal{D}(\overline{\Omega}))^n$ is dense in $H(\mathrm{div}\,;\,\Omega)$, γ_ν can be extended by continuity to a mapping still called $\gamma_\nu \in \mathcal{L}(H(\mathrm{div}\,;\,\Omega)\,;\,H^{-1/2}(\Gamma))$ such that :

$$\|\gamma_\nu\vec{v}\|_{-1/2,\Gamma} \leq \|\vec{v}\|_{H(\mathrm{div}\,;\,\Omega)} . \tag{2.6}$$ ■

By extension, $\gamma_\nu\vec{v}$ is called the normal component of $\vec{v}$ on Γ .

From Theorems 2.1 and 2.2 , we derive the next result .

<u>COROLLARY 2.1</u>.

$$(\vec{v},\overrightarrow{\mathrm{grad}}\,\varphi) + (\mathrm{div}\,\vec{v},\varphi) = \langle \gamma_\nu\vec{v},\gamma_o\varphi\rangle_\Gamma \quad \forall\, \vec{v} \in H(\mathrm{div}\,;\,\Omega),\ \forall\, \varphi \in H^1(\Omega). \tag{2.7}$$

An important byproduct of Theorem 2.2 and its Corollary is that now we can extend Green's formula for the Laplace operator to a wider range of functions.

<u>COROLLARY 2.2</u>.

<u>Let</u> $u \in H^1(\Omega)$ <u>and</u> $\Delta u \in L^2(\Omega)$. <u>Then</u> $\frac{\partial u}{\partial \nu} \in H^{-1/2}(\Gamma)$ <u>and</u>

$$(\overrightarrow{\mathrm{grad}}\,u,\overrightarrow{\mathrm{grad}}\,v) = -\,(\Delta u,v) + \langle \frac{\partial u}{\partial \nu}, \gamma_o v\rangle_\Gamma \quad \forall\, v \in H^1(\Omega) . \tag{2.8}$$

<u>PROOF</u>.

We set $\vec{w} = \overrightarrow{\mathrm{grad}}\,u \in [L^2(\Omega)]^n$. Then $\mathrm{div}\,\vec{w} = \Delta u \in L^2(\Omega)$; therefore $\vec{w} \in H(\mathrm{div}\,;\,\Omega)$ and we can apply (2.7) :

$$(\vec{w},\overrightarrow{\mathrm{grad}}\,v) + (\mathrm{div}\,\vec{w},v) = \langle \gamma_\nu\vec{w},\gamma_o v\rangle_\Gamma \quad \forall\, v \in H^1(\Omega) .$$

But $\gamma_\nu\vec{w} = \overrightarrow{\mathrm{grad}}\,u\cdot\vec{\nu}_{|\Gamma} = \frac{\partial u}{\partial \nu}$. Hence $\frac{\partial u}{\partial \nu} \in H^{-1/2}(\Gamma)$ and (2.8) is valid. ■

Another interesting consequence is that now we can interpret properly the variational problem (N') of § 1.4 and show that it is equivalent to the Neumann's problem (N).

COROLLARY 2.3.

Problems (N) and (N') of § 1.4 are equivalent.

PROOF.

We have already shown in § 1.4 that any solution u of (N') satisfies

$$-\Delta u = f \text{ in } \Omega$$

and

$$(1.31) \qquad (\overrightarrow{\text{grad}}\, u, \overrightarrow{\text{grad}}\, v) = (-\Delta u, v) + \langle g, v \rangle_\Gamma \quad \forall v \in H^1(\Omega) .$$

We must prove that $\frac{\partial u}{\partial \nu} = g$. Since $f \in L^2(\Omega)$, this follows immediately from (2.8). Hence, u is a solution of problem (N).

Conversely, according to (2.8), every solution $u \in H^1(\Omega)$ of problem (N) satisfies (1.31). ∎

Remark 2.2.

Let us go back to problem (N) in the proof of Theorem 2.1 . According to (2.7), the choice of e_i and the extension operator π, we have :

$$\langle \ell, v \rangle = \sum_{i=0}^{p} \frac{-1}{\text{meas}(\Omega_i)} \langle \gamma_\nu \vec{u}, 1 \rangle_{\Gamma_i} \int_{\Omega_i} v \, dx + \langle \gamma_\nu \vec{u}, \gamma_o v \rangle_\Gamma ,$$

where $\vec{\nu}$ points *inside* Ω. As $\gamma_\nu \vec{u} \in H^{-1/2}(\Gamma)$, it follows from Corollary 2.2 applied in $\mathcal{O} - \bar{\Omega}$ that problem (N) is equivalent to the $p+1$ non-homogeneous Neumann's problems : Find $\dot{w} \in V$ such that

$$\begin{cases} \Delta w = \dfrac{1}{\text{meas}(\Omega_i)} \langle \gamma_\nu \vec{u}, 1 \rangle_{\Gamma_i} \quad \text{in } \Omega_i , \quad 0 \leqslant i \leqslant p , \\[2ex] \dfrac{\partial w}{\partial \nu} = \gamma_\nu \vec{u} \quad \text{on } \Gamma , \quad \dfrac{\partial w}{\partial \nu} = 0 \quad \text{on } \partial\mathcal{O} . \end{cases}$$

Clearly, the compatibility condition (1.25) is satisfied in each Ω_i. Hence the extension of $\vec{u}$ has the same normal component as $\vec{u}$ on Γ. ∎

COROLLARY 2.4.

1) The range space of γ_ν is exactly $H^{-1/2}(\Gamma)$.

2) $$\| \gamma_\nu \|_{\mathcal{L}(H(\text{div}\,;\,\Omega);\, H^{-1/2}(\Gamma))} = 1 .$$

PROOF.

Let $\mu^* \in H^{-1/2}(\Gamma)$ and consider the problem :

Find φ in $H^1(\Omega)$ such that

(2.9) $$-\Delta\varphi + \varphi = 0 \quad \text{in } \Omega ,$$

(2.10) $$\frac{\partial\varphi}{\partial\nu} = \mu^* \quad \text{on } \Gamma .$$

Unlike the Neumann's problem of § 1.4, this problem has exactly one solution in $H^1(\Omega)$. We denote it by φ and set $\vec{v} = \overrightarrow{\text{grad}}\,\varphi$. Then $\vec{v} \in H(\text{div}\ ;\ \Omega)$ and $\gamma_\nu \vec{v} = \mu^*$. Moreover

$$\|\varphi\|^2_{1,\Omega} = \langle \mu^*, \gamma_o\varphi \rangle_\Gamma \leq \|\mu^*\|_{-1/2,\Gamma}\, \|\varphi\|_{1,\Omega} .$$

As $\text{div}\,\vec{v} = \varphi$, it follows that $\|\vec{v}\|_{H(\text{div}\ ;\ \Omega)} = \|\varphi\|_{1,\Omega} \leq \|\mu^*\|_{-1/2,\Gamma}$.

By (2.6), we get $\|\gamma_\nu \vec{v}\|_{-1/2,\Gamma} = \|\vec{v}\|_{H(\text{div}\ ;\ \Omega)}$. ∎

THEOREM 2.3.

$$\text{Ker}\ \gamma_\nu = H_o(\text{div}\ ;\ \Omega) .$$

The proof can be found in Duvaut & Lions [22] , and in the Appendix.

2.2. The space $H(\overrightarrow{\text{curl}}\ ;\ \Omega)$

Let us first consider the *case* $n = 2$. For $\varphi \in \mathcal{D}'(\Omega)$ and $\vec{v} \in (\mathcal{D}(\Omega))^2$, we introduce the following distributions :

(2.11) $$\overrightarrow{\text{curl}}\ \varphi = \left(\frac{\partial\varphi}{\partial x_2} , -\frac{\partial\varphi}{\partial x_1}\right)$$

and

(2.12) $$\text{curl}\ \vec{v} = \frac{\partial v_2}{\partial x_1} - \frac{\partial v_1}{\partial x_2} .$$

Then, we define the space $H(\text{curl}\ ;\ \Omega)$ as

$$H(\text{curl}\ ;\ \Omega) = \{\vec{v} \in (L^2(\Omega))^2\ ;\ \text{curl}\ \vec{v} \in L^2(\Omega)\} ,$$

a Hilbert space for the norm

(2.13) $$\|\vec{v}\|_{H(\text{curl}\ ;\ \Omega)} = \{\|\vec{v}\|^2_{o,\Omega} + \|\text{curl}\ \vec{v}\|^2_{o,\Omega}\}^{1/2} .$$

It is easy to derive many properties of $H(\text{curl}\ ;\ \Omega)$ from those of $H(\text{div}\ ;\ \Omega)$ by means of the following device : the vector $\vec{w}$ with components $(-v_2, v_1)$ belongs to H(div ; Ω) iff $\vec{v} \in$ H(curl ; Ω). Moreover, if $\vec{\tau}$ denotes the unit tangent to Γ like in figure 2 — i.e. $\vec{\tau} = (-\nu_2, \nu_1)$, then

$$\vec{w}.\vec{\nu} = -\vec{v}.\vec{\tau} \ .$$

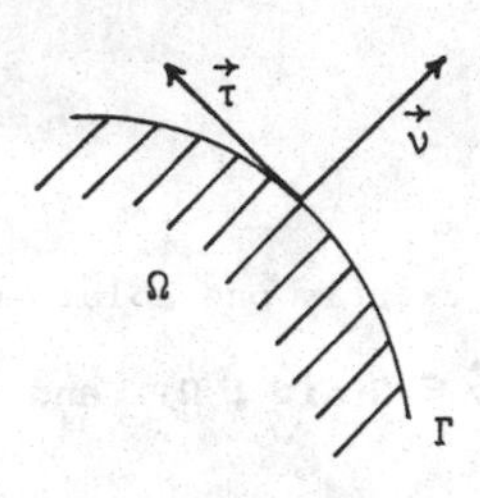

FIGURE 2

Hence the following properties stated in § 2.1 carry over to H(curl ; Ω) :

i) $(\mathcal{D}(\bar{\Omega}))^2$ is dense in H(curl ; Ω) ;

ii) the mapping γ_τ: $\vec{v} \mapsto \vec{v}\cdot\vec{\tau}|_\Gamma$ defined on $(\mathcal{D}(\bar{\Omega}))^2$ can be extended by continuity to a linear mapping still called γ_τ from H(curl ; Ω) *onto* $H^{-1/2}(\Gamma)$;

iii) H_o(curl ; Ω) defined as $\overline{(\mathcal{D}(\Omega))^2}^{\,H(\mathrm{curl}\,;\,\Omega)}$ coincides with the kernel : Ker γ_τ in H(curl ; Ω) ;

iv) the following Green's formula is valid :

$$(\vec{v}, \overrightarrow{\mathrm{curl}}\,\varphi) - (\mathrm{curl}\,\vec{v}, \varphi) = -\langle \gamma_\tau \vec{v}, \gamma_o \varphi \rangle_\Gamma \qquad \forall\, \vec{v} \in H(\mathrm{curl}\,;\,\Omega)\,, \quad \forall\, \varphi \in H^1(\Omega). \tag{2.14}$$

We now turn to the case $n = 3$. When $\vec{v} \in (\mathcal{D}'(\Omega))^3$, we define its $\overrightarrow{\mathrm{curl}}$ by

$$\overrightarrow{\mathrm{curl}}\,\vec{v} = \left(\frac{\partial v_3}{\partial x_2} - \frac{\partial v_2}{\partial x_3}\,,\ \frac{\partial v_1}{\partial x_3} - \frac{\partial v_3}{\partial x_1}\,,\ \frac{\partial v_2}{\partial x_1} - \frac{\partial v_1}{\partial x_2}\right). \tag{2.15}$$

With this operator, we introduce the following space :

$$H(\overrightarrow{\mathrm{curl}}\,;\,\Omega) = \{\vec{v} \in (L^2(\Omega))^3 \ ;\ \overrightarrow{\mathrm{curl}}\,\vec{v} \in (L^2(\Omega))^3\}$$

normed by

$$\|\vec{v}\|_{H(\overrightarrow{\mathrm{curl}}\,;\,\Omega)} = \{\|\vec{v}\|^2_{o,\Omega} + \|\overrightarrow{\mathrm{curl}}\,\vec{v}\|^2_{o,\Omega}\}^{1/2} \ . \tag{2.16}$$

<u>THEOREM 2.4.</u>

<u>The space</u> $(\mathcal{D}(\bar{\Omega}))^3$ <u>is dense in</u> H($\overrightarrow{\mathrm{curl}}$; Ω) .

This theorem is proved in Duvaut & Lions [22] .

THEOREM 2.5.

Let $\vec{v} \times \vec{\nu}$ denote the vector product of $\vec{v}$ and $\vec{\nu}$. The mapping $\vec{v} \longmapsto \vec{v} \times \vec{\nu}_{|\Gamma}$ defined on $(\mathcal{D}(\bar{\Omega}))^3$ can be extended by continuity to a linear continuous mapping still written $\vec{v} \longmapsto \vec{v} \times \vec{\nu}_{|\Gamma}$ from $H(\overrightarrow{\text{curl}}\ ;\ \Omega)$ onto $(H^{-1/2}(\Gamma))^3$.

PROOF.

The proof is similar to that of Theorem 2.2.

For all $\vec{\varphi}$ and $\vec{v}$ in $(\mathcal{D}(\bar{\Omega}))^3$, the following Green's formula is valid :

$$(\vec{\varphi}, \overrightarrow{\text{curl}}\ \vec{v}) - (\vec{v}, \overrightarrow{\text{curl}}\ \vec{\varphi}) = \int_\Gamma (\vec{v} \times \vec{\nu}) \cdot \vec{\varphi}\, d\sigma .$$

By density, this equality also holds for all $\vec{\varphi}$ in $(H^1(\Omega))^3$. Therefore

$$\left| \int_\Gamma (\vec{v} \times \vec{\nu}) . \vec{\varphi}\, d\sigma \right| \leq \|\vec{v}\|_{H(\overrightarrow{\text{curl}}\ ;\ \Omega)} \|\vec{\varphi}\|_{1,\Omega} \quad \forall\, \vec{v} \in (\mathcal{D}(\bar{\Omega}))^3 ,\ \forall\, \vec{\varphi} \in (H^1(\Omega))^3 .$$

Now, for each $\vec{\mu}$ in $(H^{1/2}(\Gamma))^3$ there exists a $\vec{\varphi}$ in $(H^1(\Omega))^3$ such that $\gamma_o \vec{\varphi} = \vec{\mu}$.

Hence $\left| \int_\Gamma (\vec{v} \times \vec{\nu}) \cdot \vec{\mu}\, d\sigma \right| \leq \|\vec{v}\|_{H(\overrightarrow{\text{curl}}\ ;\ \Omega)} \|\vec{\mu}\|_{1/2,\Gamma}$. Therefore,

$$\|\vec{v} \times \vec{\nu}\|_{-1/2,\Gamma} \leq \|\vec{v}\|_{H(\overrightarrow{\text{curl}}\ ;\ \Omega)} \quad \forall\, \vec{v} \in (\mathcal{D}(\bar{\Omega}))^3 . \tag{2.17}$$

This permits us to extend the mapping $\vec{v} \longmapsto \vec{v} \times \vec{\nu}_{|\Gamma}$ by continuity to a mapping from $H(\overrightarrow{\text{curl}}\ ;\ \Omega)$ into $(H^{-1/2}(\Gamma))^3$, satisfying (2.17) for $\vec{v}$ in $H(\overrightarrow{\text{curl}}\ ;\ \Omega)$.

In addition, Green's formula becomes :

$$(\vec{\varphi}, \overrightarrow{\text{curl}}\ \vec{v}) - (\vec{v}, \overrightarrow{\text{curl}}\ \vec{\varphi}) = \langle \vec{v} \times \vec{\nu}, \gamma_o \vec{\varphi} \rangle_\Gamma \quad \forall\, \vec{v} \in H(\overrightarrow{\text{curl}}\ ;\ \Omega),\ \vec{\varphi} \in (H^1(\Omega))^3 . \tag{2.18}$$ ■

Let us introduce the following subspace of $H(\overrightarrow{\text{curl}}\ ;\ \Omega)$:

$$H_o(\overrightarrow{\text{curl}}\ ;\ \Omega) = \overline{(\mathcal{D}(\Omega))^3}^{H(\overrightarrow{\text{curl}}\ ;\ \Omega)} .$$

THEOREM 2.6.

We have

$$H_o(\overrightarrow{\text{curl}}\ ;\ \Omega) = \{\vec{v} \in H(\overrightarrow{\text{curl}}\ ;\ \Omega)\ ;\ \vec{v} \times \vec{\nu}_{|\Gamma} = \vec{0}\} .$$

Remark 2.3.

If $\vec{v} \in H_o(\overrightarrow{\text{curl}}\ ;\ \Omega)$, its tangential components vanish on Γ. ■

§ 3 - A DECOMPOSITION OF VECTOR FIELDS

In this paragraph, we shall prove that every vector of $(L^2(\Omega))^n$ is the sum of a divergence-free vector and a gradient vector. This will lead to an interesting decomposition of $(L^2(\Omega))^n$ and $(H_0^1(\Omega))^n$ as a direct sum of orthogonal spaces.

We shall make the following assumptions on Ω : Ω is bounded, eventually multiply-connected, and its boundary Γ is Lipschitz continuous.

Like in § 2, we shall denote by Γ_o the exterior boundary of Ω and by Γ_i , $1 \leq i \leq p$, the other components of Γ (cf. figure 1). The duality between $H^{-1/2}(\Gamma_i)$ and $H^{1/2}(\Gamma_i)$ will be denoted by $< \cdot,\cdot >_{\Gamma_i}$.

3.1. Existence of the stream function of a divergence-free vector

We first consider the case $n = 2$.

THEOREM 3.1.

Let $n = 2$. A function $\vec{v} \in [L^2(\Omega)]^2$ satisfies

$$\text{(3.1)} \qquad \operatorname{div} \vec{v} = 0 \quad , \quad < \gamma_\nu \vec{v}, 1 >_{\Gamma_i} = 0 \quad \text{for} \quad 0 \leq i \leq p$$

if and only if there exists a stream function φ in $H^1(\Omega)$ such that

$$\text{(3.2)} \qquad \vec{v} = \overrightarrow{\operatorname{curl}}\, \varphi \; .$$

Proof.

1) Let us show that (3.2) implies (3.1). Let $\varphi \in H^1(\Omega)$ and let $\vec{v} = \overrightarrow{\operatorname{curl}}\, \varphi$; then

$$\operatorname{div}(\overrightarrow{\operatorname{curl}}\, \varphi) = 0 \; .$$

Next, as $\mathcal{D}(\overline{\Omega})$ is dense in $H^1(\Omega)$, it suffices to prove that

$$\int_{\Gamma_i} \overrightarrow{\operatorname{curl}}\, \varphi \cdot \vec{\nu} \, d\sigma = 0 \qquad \forall\, \varphi \in \mathcal{D}(\overline{\Omega}) \;, \quad 0 \leq i \leq p \; .$$

But

$$\int_{\Gamma_i} \overrightarrow{\operatorname{curl}}\, \varphi \cdot \vec{\nu} \, d\sigma = \int_{\Gamma_i} \frac{\partial \varphi}{\partial \tau} \, d\sigma = 0 \; .$$

Therefore (3.2) implies (3.1).

2) Conversely, let $\vec{v}$ satisfy (3.1). The idea is to extend $\vec{v}$ to the whole plane in such a way that it stays divergence-free. Then it will be easy to construct its stream function by means of Fourier transforms. The extension procedure is similar to that used in the proof of Theorem 2.1.

a) Let $\mathcal{O}$ be any bounded, simply-connected open set containing Ω, i.e. $\overline{\Omega} \subset \mathcal{O}$. Then, for $p \geq 1$, the set $\mathcal{O}-\overline{\Omega}$ is not connected and again we call Ω_i that component which is bounded by Γ_i, for $1 \leq i \leq p$, and Ω_o that component bounded by Γ_o and $\partial\mathcal{O}$. Consider the following problem.

Find a function w defined in $\mathcal{O}-\overline{\Omega}$ and such that

$$\text{(N)} \quad \begin{cases} \Delta w = 0 \quad \text{in} \quad \mathcal{O}-\overline{\Omega}, \\ \dfrac{\partial w}{\partial \nu} = \vec{v}\cdot\vec{\nu} \quad \text{on} \quad \Gamma_i \quad \text{for} \quad 0 \leq i \leq p, \\ \dfrac{\partial w}{\partial \nu} = 0 \quad \text{on} \quad \partial\mathcal{O}. \end{cases}$$

Here again, problem (N) consists of $p+1$ non-homogeneous Neumann's problems (in Ω_i, $0 \leq i \leq p$) like the one we analyzed in § 1.4, since they include the compatibility conditions :

$$< \vec{v}.\vec{\nu}, 1 >_{\Gamma_i} = 0 \quad \text{for} \quad 0 \leq i \leq p .$$

Therefore, there exists a function $w \in H^1(\mathcal{O}-\overline{\Omega})$, determined uniquely up to an additive constant in each Ω_i, satisfying (N). We set

$$\text{(3.3)} \qquad \vec{\theta} = \overrightarrow{\text{grad}}\, w .$$

Then $\vec{\theta} \in H(\text{div}\,; \mathcal{O}-\overline{\Omega})$,

$$\text{div}\,\vec{\theta} = \Delta w = 0 \quad \text{in} \quad \mathcal{O}-\overline{\Omega},$$

$$\text{(3.4)} \quad \begin{cases} \vec{\theta}\cdot\vec{\nu} = \dfrac{\partial w}{\partial \nu} = \vec{v}\cdot\vec{\nu} \quad \text{on each} \quad \Gamma_i, \ 0 \leq i \leq p, \\ \vec{\theta}\cdot\vec{\nu} = 0 \quad \text{on} \quad \partial\mathcal{O}. \end{cases}$$

b) Now, we extend $\vec{v}$ as follows :

$$\tilde{\vec{v}} = \begin{cases} \vec{v} \quad \text{in} \quad \Omega, \\ \vec{\theta} \quad \text{in} \quad \mathcal{O}-\overline{\Omega}, \\ \vec{0} \quad \text{elsewhere}. \end{cases}$$

Clearly $\tilde{\vec{v}} \in (L^2(\mathbb{R}^2))^2$. Let us calculate its divergence. As a distribution, $\operatorname{div} \tilde{\vec{v}}$ satisfies

$$< \operatorname{div} \tilde{\vec{v}}, \varphi > = - \int_{\mathbb{R}^2} \tilde{\vec{v}} \cdot \overrightarrow{\operatorname{grad}}\, \varphi \, dx \qquad \forall \varphi \in \mathcal{D}(\mathbb{R}^2) \quad ,$$

that is

$$< \operatorname{div} \tilde{\vec{v}}, \varphi > \; = - \int_{\Omega} \vec{v} . \overrightarrow{\operatorname{grad}}\, \varphi \, dx - \sum_{i=0}^{p} \int_{\Omega_i} \vec{\theta} \cdot \overrightarrow{\operatorname{grad}}\, \varphi \, dx \; .$$

As $\vec{v}$ and $\vec{\theta}$ are both divergence-free, Green's formula (2.7) and (3.4) yield

(3.5)
$$< \operatorname{div} \tilde{\vec{v}}, \varphi > = - \langle \gamma_\nu \vec{v}, \gamma_o \varphi >_{\Gamma} - \sum_{i=0}^{p} < \gamma_\nu \vec{v}, \gamma_o \varphi >_{\Gamma_i} \; .$$

In the sum, the normal to Γ_i is directed outside Ω_i and therefore *inside* Ω .
Hence each term of this sum cancels a term of $< \gamma_\nu \vec{v}, \gamma_o \varphi >_\Gamma$. Therefore

$$< \operatorname{div} \tilde{\vec{v}}, \varphi > = 0 \qquad \forall \varphi \in \mathcal{D}(\mathbb{R}^2) \; .$$

Therefore

$$\tilde{\vec{v}} \in H(\operatorname{div} ; \mathbb{R}^2) \; ,$$
$$\operatorname{div} \tilde{\vec{v}} = 0 \; .$$

c) Let us introduce the Fourier transform of $\tilde{\vec{v}}$:

(3.6) $\hat{\tilde{v}}_j(\xi) = \int_{\mathbb{R}^2} e^{-2i\pi(x,\xi)} \tilde{v}_j(x) \, dx$, $j = 1,2$, where $(x,\xi) = x_1\xi_1 + x_2\xi_2$.

Note that each $\hat{\tilde{v}}_j(\xi)$ is a holomorphic function of the complex variables ξ_1 and ξ_2 since the support of $\tilde{v}_j$ is compact.
In terms of Fourier transforms, the condition $\operatorname{div} \tilde{\vec{v}} = 0$ becomes

(3.7)
$$\xi_1 \hat{\tilde{v}}_1 + \xi_2 \hat{\tilde{v}}_2 = 0$$

and

$$\tilde{\vec{v}} = \overrightarrow{\operatorname{curl}}\, \varphi$$

holds if and only if

(3.8)
$$\hat{\tilde{v}}_1 = 2i\pi\xi_2 \hat{\varphi} \; , \quad \hat{\tilde{v}}_2 = - 2i\pi\xi_1 \hat{\varphi} \quad .$$

Now, if we take $\hat{\varphi} = \frac{1}{2i\pi\xi_2} \hat{\tilde{v}}_1$ then, thanks to (3.7) both equalities of (3.8) are valid, i.e. $\hat{\varphi} = \frac{\hat{\tilde{v}}_1}{2i\pi\xi_2} = \frac{-\hat{\tilde{v}}_2}{2i\pi\xi_1}$. Therefore, the inverse transform of $\hat{\varphi}$ is the required stream function of $\tilde{\vec{v}}$, provided $\hat{\varphi} \in L^2(\mathbb{R}^2_\xi)$. As $\hat{\tilde{v}}_j \in L^2(\mathbb{R}^2_\xi)$, it suffices to show that $\hat{\varphi}$ is bounded in a neighborhood of the origin.

According to (3.7), we have $\tilde{\hat{v}}_1(\xi_1,0) = 0$. Hence, using the holomorphy of the function $\hat{v}_1$, we obtain :

$$\tilde{\hat{v}}_1(\xi_1,\xi_2) = \xi_2 \frac{\partial \tilde{\hat{v}}_1}{\partial \xi_2}(\xi_1,0) + O(|\xi_2|^2) ,$$

so that

$$\hat{\varphi}(\xi_1,\xi_2) = \frac{1}{2i\pi} \frac{\partial \tilde{\hat{v}}_1}{\partial \xi_2}(\xi_1,0) + O(|\xi_2|) .$$

Clearly, this implies that $\hat{\varphi}$ is bounded in a neighborhood of zero . ∎

Remark 3.1.

If Ω is connected, then clearly the stream function φ of $\vec{v}$ is unique up to an additive constant. ∎

Suppose again that Ω is connected (otherwise, we deal with each of its components separately) and consider the space

$$\mathbb{H} = \{\vec{v} \in H_o(\text{div} ; \Omega) ; \text{div } \vec{v} = 0\} .$$

Then every stream function φ of $\vec{v}$ satisfies $\frac{\partial \varphi}{\partial \tau}\Big|_{\Gamma_i} = 0$, that is :

$$\varphi|_{\Gamma_i} = \text{a constant } c_i \quad , \quad \text{for } 0 \leqslant i \leqslant p .$$

According to the above remark, φ is uniquely determined if we fix one of these constants. Therefore, $\vec{v}$ has one and only one stream function φ that *vanishes on* Γ_o. Let us characterize this function as the solution of a boundary value problem. For this, we introduce the space

$$\tilde{\Phi} = \{\chi \in H^1(\Omega) ; \chi|_{\Gamma_o} = 0 \quad , \chi|_{\Gamma_i} = \text{an arbitrary constant} \quad \text{for } 1 \leqslant i \leqslant p\},$$

which is a closed subspace of $H^1(\Omega)$. Moreover, $|\cdot|_{1,\Omega}$ and $\|\cdot\|_{1,\Omega}$ are two equivalent norms on $\tilde{\Phi}$, by virtue of the following generalization of the Poincaré-Friedrichs' inequality :

LEMMA 3.1.

Let Ω be a bounded and connected open subset of $\mathbb{R}^n$, with a Lipschitz continuous boundary Γ . Let $\Gamma_o \subset \Gamma$ with meas$(\Gamma_o) > 0$. Then, $|\cdot|_{1,\Omega}$ and $\|\cdot\|_{1,\Omega}$ are two equivalent norms on the space $\{v \in H^1(\Omega) ; v|_{\Gamma_o} = 0\}$.

Then, the stream function of $\vec{v}$ that vanishes on Γ_0 is also the only solution of the problem :

Find φ in $\tilde{\Phi}$ such that

$$(\vec{\text{curl}}\, \varphi, \vec{\text{curl}}\, \chi) = (\vec{v}, \vec{\text{curl}}\, \chi) \quad \forall\, \chi \in \tilde{\Phi} . \tag{3.9}$$

Let us interpret (3.9). If we restrict χ to $\mathcal{D}(\Omega)$, we get :

$$- \Delta\varphi = \text{curl}\, \vec{v} \quad \text{in} \quad \mathcal{D}'(\Omega) .$$

Next, by applying *formally* (2.8) to φ , we obtain :

$$(\vec{\text{curl}}\, \varphi, \vec{\text{curl}}\, \chi) = - (\Delta\varphi, \chi) + < \frac{\partial\varphi}{\partial\nu}, \gamma_0\chi >_\Gamma \quad \forall\, \chi \in \tilde{\Phi} .$$

Likewise, (2.14) yields formally

$$(\vec{v}, \vec{\text{curl}}\, \chi) = (\text{curl}\, \vec{v}, \chi) - < \gamma_\tau\vec{v}, \gamma_0\chi >_\Gamma \quad , \quad \forall\, \chi \in \tilde{\Phi} .$$

Finally, by comparing these two equalities with (3.9) , we get :

$$\sum_{i=1}^{p} < \frac{\partial\varphi}{\partial\nu} + \gamma_\tau\vec{v}, c_i >_{\Gamma_i} = 0 \quad \forall\, c_i \in \mathbb{R} .$$

These results are summarized in the following corollary.

COROLLARY 3.1.

Let Ω be like in Lemma 3.1. For $\vec{v} \in \mathbf{H}$, the relation :

$$\vec{v} = \vec{\text{curl}}\, \varphi$$

establishes a one-to-one correspondence between the spaces $\mathbf{H}$ and $\tilde{\Phi}$. Furthermore, the stream function φ can be characterized as the solution of problem (3.9). This problem has the following interpretation :

$$- \Delta\varphi = \text{curl}\, \vec{v} \quad \text{in} \quad \mathcal{D}'(\Omega) ,$$

$$\varphi|_{\Gamma_0} = 0 \quad , \quad \varphi|_{\Gamma_i} = c_i \quad , \quad 1 \leq i \leq p ,$$

where the constants c_i are determined *formally* by :

$$< \frac{\partial\varphi}{\partial\nu} + \gamma_\tau\vec{v}, 1 >_{\Gamma_i} = 0 \quad , \qquad 1 \leq i \leq p .$$

Now, we turn to the case $n = 3$.

THEOREM 3.2.

Let $n = 3$. *A function* $\vec{v} \in (L^2(\Omega))^3$ *satisfies* :

(3.10) $$\operatorname{div} \vec{v} = 0 \text{ in } \Omega \,,\quad < \vec{v}\cdot\vec{\nu}, 1 >_{\Gamma_i} = 0 \,,\text{ for } 0 \leq i \leq p$$

if and only if there exists a function $\vec{\varphi}$ *in* $(H^1(\Omega))^3$ *such that*

(3.11) $$\vec{v} = \vec{\operatorname{curl}}\, \vec{\varphi} \,.$$

PROOF.

1) Let $\vec{\varphi}$ belong to $(H^1(\Omega))^3$ and $\vec{v} = \vec{\operatorname{curl}}\, \vec{\varphi}$. Then $\operatorname{div} \vec{v} = 0$, and we must check that $< \vec{v}\cdot\vec{\nu}, 1 >_{\Gamma_i} = 0$. For $0 \leq i \leq p$, let θ_i be a function of $\mathcal{D}(\mathbb{R}^3)$ such that

$$0 \leq \theta_i(x) \leq 1 \text{ in } \mathbb{R}^3 \text{ and } \theta_i(x) \equiv \delta_i^j \text{ in a neighborhood of } \Gamma_j \,.$$

We set $\vec{w}_i = \vec{\operatorname{curl}}(\theta_i \vec{\varphi})$. Obviously, $\vec{w}_i \in [L^2(\Omega)]^3$ and $\operatorname{div} \vec{w}_i = 0$.
Moreover,

$$\vec{w}_i\cdot\vec{\nu}|_{\Gamma_j} = 0 \quad\text{if}\quad j \neq i \quad,\quad \vec{w}_i\cdot\vec{\nu}|_{\Gamma_i} = \vec{v}\cdot\vec{\nu}|_{\Gamma_i} \,.$$

Hence

$$< \vec{v}\cdot\vec{\nu}, 1 >_{\Gamma_i} = < \vec{w}_i\cdot\vec{\nu}, 1 >_{\Gamma} = \int_\Omega \operatorname{div} \vec{w}_i \, dx = 0 \,,\ \text{for}\quad 0 \leq i \leq p.$$

2) Conversely, let $\vec{v}$ be a function of $[L^2(\Omega)]^3$ satisfying (3.10). As in the two-dimensional case, we can extend $\vec{v}$ to the whole space so that the extended function $\vec{\tilde{v}} \in [L^2(\mathbb{R}^3)]^3$ is divergence-free and has a compact support. Again, let $\hat{v}_j$ be the Fourier transform of $\tilde{v}_j$:

$$\hat{v}_j(\xi) = \int_{\mathbb{R}^3} e^{-2i\pi(x,\xi)} \tilde{v}_j(x)\, dx \,.$$

Then $\hat{v}_j$ is holomorphic in $\mathbb{R}^3_\xi$ since the support of $\tilde{v}_j$ is compact. The condition $\operatorname{div} \vec{\tilde{v}} = 0$ becomes :

(3.12) $$\sum_{i=1}^{3} \xi_i \hat{v}_i = 0 \,.$$

Now, we must find a function $\vec{\hat{\varphi}}$ in $[L^2(\mathbb{R}^3_\xi)]^3$ such that $\vec{\operatorname{curl}}\, \vec{\varphi} = \vec{\tilde{v}}$, i.e. such that :

$$(3.13)\quad \begin{cases} \hat{v}_1 = 2i\pi(\xi_2\hat{\varphi}_3 - \xi_3\hat{\varphi}_2)\ , \\ \hat{v}_2 = 2i\pi(\xi_3\hat{\varphi}_1 - \xi_1\hat{\varphi}_3)\ , \\ \hat{v}_3 = 2i\pi(\xi_1\hat{\varphi}_2 - \xi_2\hat{\varphi}_1)\ . \end{cases}$$

The third equation of (3.13) is a consequence of the first two and (3.12). Therefore, in order to fix $\hat{\varphi}$, we add to (3.13) the condition

$$(3.14)\quad \sum_{i=1}^{3} \xi_i\hat{\varphi}_i = 0\ .$$

The unique solution of (3.12), (3.13) and (3.14) is :

$$\hat{\varphi}_1 = \frac{1}{2i\pi\, \|\xi\|^2}(\xi_3\hat{v}_2 - \xi_2\hat{v}_3)\ ,$$

$$\hat{\varphi}_2 = \frac{1}{2i\pi\, \|\xi\|^2}(\xi_1\hat{v}_3 - \xi_3\hat{v}_1)$$

and

$$\hat{\varphi}_3 = \frac{1}{2i\pi\, \|\xi\|^2}(\xi_2\hat{v}_1 - \xi_1\hat{v}_2)\ ,$$

where $\|\xi\|^2 = \sum_{i=1}^{3} \xi_i^2$. By inspection, these equations imply that $\xi_j\hat{\varphi}_i \in L^2(\mathbb{R}_\xi)$ and that $|\hat{\varphi}_i| \le \frac{1}{2\pi\,\|\xi\|}(|\hat{v}_j| + |\hat{v}_k|)$. Therefore, the inverse transform of $\hat{\vec{\varphi}}$ belongs to $(H^1(\Omega))^3$, provided that $\hat{\vec{\varphi}}$ is bounded at the origin. First, we observe that (3.12) implies that $\hat{v}_i(0) = 0$. Then, as $\hat{v}_i$ is holomorphic, it follows that $\hat{v}_i(\xi) = \sum_{j=1}^{3} \xi_j \frac{\partial \hat{v}_i}{\partial \xi_j}(0) + O(\|\xi\|^2)$ in a neighborhood of 0 . Hence $\hat{\vec{\varphi}}$ is bounded as ξ tends to zero.

By restricting to Ω the inverse transform $\vec{\varphi}$ of $\hat{\vec{\varphi}}$, we thus find a function $\vec{\varphi}$ in $(H^1(\Omega))^3$ such that $\vec{\text{curl}}\,\vec{\varphi} = \vec{v}$ and moreover, $\text{div}\,\vec{\varphi} = 0$, which is slightly stronger than the statement of the theorem. ∎

<u>Remarks 3.2.</u>

1) The divergence-free stream function $\vec{\varphi}$ of $\tilde{\vec{v}}$ is unique, but this does not hold for $\vec{v}$, since the extension of $\vec{v}$ is not unique.

2) From the identity in $(\mathcal{D}'(\Omega))^3$:

$$(3.15)\quad \vec{\text{curl}}(\vec{\text{curl}}\,\vec{\varphi}) = -\Delta\vec{\varphi} + \vec{\text{grad}}\,\text{div}\,\vec{\varphi}\ ,$$

we see that $\vec{\varphi}$ satisfies

$$- \Delta\vec{\varphi} = \vec{\text{curl}}\ \vec{v} \ . \tag{3.16}$$

The choice of the boundary conditions that must be added to (3.16) in order to characterize $\vec{\varphi}$ can be found in Bernardi [8].

3) In the above proof, we have chosen a divergence-free stream function, but of course this is not the only possibility. For instance, we might have replaced (3.14) by

$$\hat{\varphi}_3 = 0 \ . \tag{3.14'}$$

Then (3.12), (3.13) and (3.14') have the only solution

$$\hat{\varphi}_1 = \frac{\hat{v}_2}{2i\pi\xi_3} \ , \quad \hat{\varphi}_2 = -\frac{\hat{v}_1}{2i\pi\xi_3} \ , \quad \hat{\varphi}_3 = 0 \ .$$

The corresponding stream function $\vec{\varphi}$ of $\vec{v}$ has the form $(\varphi_1, \varphi_2, 0)$. ■

3.2. A decomposition of $[L^2(\Omega)]^n$

THEOREM 3.3

For each function $\vec{v}$ of $[L^2(\Omega)]^n$, there exists a function $q \in H^1(\Omega)$ and a function $\varphi \in H^1(\Omega)$ if $n = 2$ (respectively, $\vec{\varphi} \in (H^1(\Omega))^3$ if $n = 3$), such that

$$\begin{cases} \vec{v} = \vec{\text{grad}}\ q + \begin{cases} \vec{\text{curl}}\ \varphi & \text{if} \quad n = 2 \\ \vec{\text{curl}}\ \vec{\varphi} & \text{if} \quad n = 3 \ , \end{cases} \\ \gamma_\nu(\vec{v} - \vec{\text{grad}}\ q) = 0 \ . \end{cases} \tag{3.17}$$

Moreover, the functions $\vec{\text{curl}}\ \varphi$ (resp. $\vec{\text{curl}}\ \vec{\varphi}$) and $\vec{\text{grad}}\ q$ are orthogonal in $(L^2(\Omega))^n$.

PROOF.

Let $\vec{v} \in [L^2(\Omega)]^n$ and consider the following problem :

find q in $H^1(\Omega)$ satisfying

$$(\vec{\text{grad}}\ q, \vec{\text{grad}}\ \mu) = (\vec{v}, \vec{\text{grad}}\ \mu) \qquad \forall\ \mu \in H^1(\Omega) \ . \tag{3.18}$$

This Neumann's problem has a unique solution $\dot{q}$ in $H^1(\Omega)/\mathbb{R}$. Any $q \in \dot{q}$ verifies

$$\Delta q = \text{div}\ \vec{v} \quad \text{in} \quad H^{-1}(\Omega) \ .$$

Hence $\vec{v} - \vec{\text{grad}}\ q$ is a divergence-free vector of $H(\text{div}\ ;\ \Omega)$. Then, by applying

Green's formula (2.7) to (3.18), we get

$$0 = (\vec{v}-\overrightarrow{\text{grad}}\ q, \overrightarrow{\text{grad}}\ \mu) = < (\vec{v}-\overrightarrow{\text{grad}}\ q).\vec{\nu},\mu >_\Gamma \qquad \forall\ \mu \in H^1(\Omega)\ .$$

This implies that $(\vec{v}-\overrightarrow{\text{grad}}\ q)\cdot\vec{\nu}_{|\Gamma} = 0$ in $H^{-1/2}(\Gamma)$. Therefore, we can apply to $\vec{v} - \overrightarrow{\text{grad}}\ q$ the conclusion of Theorem 3.1 (or 3.2) and we find the required function φ (or $\vec{\varphi}$). Note that this particular φ (or $\vec{\varphi}$) satisfies

$$(3.19) \qquad \gamma_\nu(\overrightarrow{\text{curl}}\ \vec{\varphi}) = 0\ .$$

To show the orthogonality of $\overrightarrow{\text{grad}}\ r$ and $\overrightarrow{\text{curl}}\ \vec{\varphi}$, we use Green's formula and (3.19)

$$(\overrightarrow{\text{grad}}\ r, \overrightarrow{\text{curl}}\ \vec{\varphi}) = -\ (r, \text{div}\ (\overrightarrow{\text{curl}}\ \vec{\varphi})) + < \gamma_\nu(\overrightarrow{\text{curl}}\ \vec{\varphi}), \gamma_0 r >_\Gamma$$
$$= 0 \qquad \forall\ r \in H^1(\Omega)\ . \qquad \blacksquare$$

<u>Remarks 3.3.</u>

1) The decomposition of Theorem 3.3 determines a unique q , up to an additive constant. The function q is characterized as the solution of (3.18) ; that is , if $\vec{v} \in H(\text{div}\ ;\ \Omega)$, then

$$\Delta q = \text{div}\ \vec{v} \quad \text{in} \quad \Omega$$

and

$$\frac{\partial q}{\partial \nu} = \vec{v}.\vec{\nu} \quad \text{on} \quad \Gamma\ .$$

If $\vec{v}$ is only in $[L^2(\Omega)]^n$, the last equality is formal.

2) When n = 2, the above decomposition and the condition $\varphi_{|\Gamma_0} = 0$ determine φ uniquely. We can characterize φ as being the only function of $\tilde{\Phi}$ that satisfies

$$(\overrightarrow{\text{curl}}\ \varphi, \overrightarrow{\text{curl}}\ \chi) = (\vec{v} - \overrightarrow{\text{grad}}\ q, \overrightarrow{\text{curl}}\ \chi) \qquad \forall\ \chi \in \tilde{\Phi}\ .$$

Hence, formally φ is the solution of

$$\begin{cases} -\ \Delta\varphi = \text{curl}\ \vec{v} \quad \text{in} \quad \Omega\ , \\ \varphi_{|\Gamma_0} = 0 \quad , \quad \varphi_{|\Gamma_i} \text{ is constant} \quad \text{for} \quad 1 \leqslant i \leqslant p\ , \\ \int_{\Gamma_i} (\frac{\partial\varphi}{\partial\nu} + \vec{v}\cdot\vec{\tau} - \frac{\partial q}{\partial\tau})d\sigma = 0 \quad , \quad \text{for} \quad 1 \leqslant i \leqslant p\ . \end{cases} \qquad \blacksquare$$

We now come to the decomposition of $(L^2(\Omega))^n$ into orthogonal subspaces. Let us introduce the space

$$\mathcal{V} = \{\vec{v} \in (\mathcal{D}(\Omega))^n \; ; \; \operatorname{div} \vec{v} = 0\} \quad ,$$

and let $H^\perp$ denote the orthogonal complement of H in $(L^2(\Omega))^n$. The next theorem characterizes H and $H^\perp$.

THEOREM 3.4.

We have

$$(L^2(\Omega))^n = H \oplus H^\perp$$

with

$$(3.20) \quad \begin{cases} H = \{\vec{\operatorname{curl}}\, \varphi \; ; \; \varphi \in (H^1(\Omega))^2 \text{ and } \gamma_\nu \, \vec{\operatorname{curl}}\, \varphi = 0\} \text{ if } n=2, \\ H = \{\vec{\operatorname{curl}}\, \vec{\varphi} \; ; \; \vec{\varphi} \in (H^1(\Omega))^3 \text{ and } \gamma_\nu \, \vec{\operatorname{curl}}\, \vec{\varphi} = 0\} \text{ if } n=3, \end{cases}$$

$$(3.21) \quad H^\perp = \{\overrightarrow{\operatorname{grad}}\, q \; ; \; q \in H^1(\Omega)\} .$$

Moreover, $\mathcal{V}$ is dense in H.

PROOF.

1) Since H is a closed subspace of $(L^2(\Omega))^n$, it follows that $(L^2(\Omega))^n = H \oplus H^\perp$. Next, the characterization (3.20) is a direct consequence of Theorems 3.1 and 3.2. Finally, if $\vec{v} \in (L^2(\Omega))^n$, then Theorem 3.3 implies that $\vec{v}$ is of the form :

$$\vec{v} = \vec{w} + \overrightarrow{\operatorname{grad}}\, q \text{ with } \vec{w} \text{ in } H .$$

Therefore $H^\perp$ is the space (3.21).

2) Let us check the density of $\mathcal{V}$ in H in the two-dimensional case. Here, we make use of Corollary 3.1 :

$$\vec{v} \in H \text{ iff } \vec{v} = \vec{\operatorname{curl}}\, \varphi \text{ with } \varphi \text{ in } \tilde{\Phi} \text{, and therefore}$$

$$\varphi|_{\Gamma_o} = 0 \text{ and } \varphi|_{\Gamma_i} = c_i \text{ for } 1 \leq i \leq p .$$

It is easy to construct a function ψ in $\mathcal{D}(\bar{\Omega})$ such that

$$\psi \equiv 0 \text{ in a neighborhood of } \Gamma_o$$

and

$$\psi \equiv c_i \text{ in a neighborhood of } \Gamma_i .$$

As $\varphi - \psi \in H^1_o(\Omega)$, there exists a sequence θ_m in $\mathcal{D}(\Omega)$ such that

$\lim_{m\to\infty} \theta_m = \varphi - \psi$ in $H^1(\Omega)$. Also $\vec{\text{curl}}\, \theta_m \in \mathcal{V}$ and clearly $\vec{\text{curl}}\, \psi \in \mathcal{V}$.

Hence
$$\vec{v} = \lim_{m\to\infty} (\vec{\text{curl}}\, \theta_m + \vec{\text{curl}}\, \psi) .$$

The proof for the three-dimensional case can be found in Temam [44]. ■

3.3. A decomposition of $(H_o^1(\Omega))^n$

In solving the Stokes problem, we shall deal with divergence-free functions of $(H^1(\Omega))^n$. The following theorem proves a boundary value property of these functions.

THEOREM 3.5.

Let $\vec{g}$ be a given function of $(\dot{H}^{1/2}(\Gamma))^n$ that satisfies $\int_\Gamma \vec{g}\cdot\vec{\nu}\, d\sigma = 0$. *Then there exists a function $\vec{u}$ in $(H^1(\Omega))^n$ such that* :

$$\text{div}\, \vec{u} = 0 \quad , \quad \vec{u} = \vec{g} \ \text{ on } \ \Gamma .$$

PROOF.

For the sake of simplicity, we assume that $n = 2$ and that Γ is sufficiently smooth. The proof for the general case can be found in Temam [44].

1) Consider first the case where $\vec{g}\cdot\vec{\nu} = 0$ on Γ . It suffices to find ψ in $H^2(\Omega)$ such that

$$\frac{\partial\psi}{\partial\tau} = 0 \ \text{ on } \ \Gamma \ , \quad \frac{\partial\psi}{\partial\nu} = - \vec{g}\cdot\vec{\tau} \ \text{ on } \ \Gamma . \tag{3.22}$$

Indeed, if we set $\vec{u} = \vec{\text{curl}}\, \psi$, then $\vec{u} \in (H^1(\Omega))^2$, $\text{div}\, \vec{u} = 0$,

$$\vec{u}\cdot\vec{\nu} = \frac{\partial\psi}{\partial\tau} = 0 = \vec{g}\cdot\vec{\nu} \quad \text{on } \Gamma ,$$

and

$$\vec{u}\cdot\vec{\tau} = - \frac{\partial\psi}{\partial\nu} = \vec{g}\cdot\vec{\tau} \quad \text{on } \Gamma ,$$

hence $\vec{u}$ is the required function .

According to Theorem 1.5 , and since $\vec{g}\cdot\vec{\tau} \in H^{1/2}(\Gamma)$, there exists a function ψ in $H^2(\Omega)$ such that $\psi = 0$ on Γ and $\frac{\partial\psi}{\partial\nu} = - \vec{g}\cdot\vec{\tau}$ on Γ .

Thus ψ satisfies (3.22).

2) When $\vec{g}\cdot\vec{\nu}$ does not vanish on Γ, we cannot apply directly Theorem 1.5 because, in general, $\vec{g}\cdot\vec{\nu} \notin H^{3/2}(\Gamma)$. Instead, we introduce the function p satisfying the Neumann's problem :

$$(3.23)\quad \begin{cases} \Delta p = 0 \quad \text{in} \quad \Omega , \\ \dfrac{\partial p}{\partial \nu} = \vec{g}\cdot\vec{\nu} \quad \text{on} \quad \Gamma . \end{cases}$$

Because of the hypothesis $< \vec{g}\cdot\vec{\nu}, 1 >_{\Gamma} = 0$, problem (3.23) has a unique solution $\dot{p}$ in $H^1(\Omega)/\mathbb{R}$. Moreover, since $\vec{g}\cdot\vec{\nu} \in H^{1/2}(\Gamma)$ then $\dot{p} \in H^2(\Omega)/\mathbb{R}$, provided that Γ *is sufficiently smooth* (cf. remark 1.2).

Then, according to part 1, there exists a function $\vec{u}_1$ in $(H^1(\Omega))^2$ such that $\operatorname{div} \vec{u}_1 = 0$ and $\vec{u}_1 = \vec{g} - \gamma_o(\overrightarrow{\text{grad}}\, p)$ on Γ.

Therefore $\vec{u} = \vec{u}_1 + \overrightarrow{\text{grad}}\, p$ is the required function. ■

Remark 3.4.

Let us assume that Γ is infinitely differentiable and that $\vec{g}$ belongs to $(H^{m-1/2}(\Gamma))^n$, for $m \geq 1$, with $\int_{\Gamma} \vec{g}\cdot\vec{\nu}\, d\sigma = 0$. Since Theorem 1.5 and remark 1.2 are both valid for any m, we can apply the above reasoning to show that there exists $\vec{u}$ in $(H^m(\Omega))^n$ such that

$$\operatorname{div} \vec{u} = 0 \quad \text{and} \quad \gamma_o \vec{u} = \vec{g} . \qquad ■$$

Now, we define the following spaces :

$$V = \{\vec{v} \in (H^1_o(\Omega))^n \ ; \ \operatorname{div} \vec{v} = 0\} ,$$

$$L^2_o(\Omega) = \{q \in L^2(\Omega) \ ; \ (q,1) = 0\}$$

and we denote by $V^{\perp}$ the orthogonal complement of V in $(H^1_o(\Omega))^n$ for the scalar product $(\overrightarrow{\text{grad}}\, \vec{u}, \overrightarrow{\text{grad}}\, \vec{v})$.

LEMMA 3.2.

<u>The divergence operator is an isomorphism from $V^{\perp}$ onto $L^2_o(\Omega)$</u> .

Proof.

Let $\vec{v} \in (H^1_o(\Omega))^n$. By Green's formula :

$$\int_\Omega \operatorname{div} \vec{v}\, dx = \int_\Gamma \vec{v}\cdot\vec{\nu}\, d\sigma = 0 .$$

Thus $\operatorname{div} \in \mathcal{L}((H_o^1(\Omega))^n ; L_o^2(\Omega))$. Let us show that div is a one-to-one mapping from $\mathbf{V}^\perp$ onto $L_o^2(\Omega)$. Since $\mathbf{V} = \operatorname{Ker}(\operatorname{div})$, it suffices to show that div maps $(H_o^1(\Omega))^n$ onto $L_o^2(\Omega)$. For this, let q be a function of $L_o^2(\Omega)$; we seek $\vec{v}$ in $(H_o^1(\Omega))^n$ such that $\operatorname{div} \vec{v} = q$. As Ω is bounded, there exists some function θ in $H^2(\Omega)$ such that

$$\Delta\theta = q \text{ in } \Omega .$$

We set $\vec{v}_1 = \overrightarrow{\operatorname{grad}}\, \theta \in (H^1(\Omega))^n$. Then

$$\operatorname{div} \vec{v}_1 = \Delta\theta = q ;$$

moreover, by Green's formula

$$\int_\Gamma \vec{v}_1\cdot\vec{\nu}\, d\sigma = \int_\Omega \operatorname{div} \vec{v}_1\, dx = \int_\Omega q\, dx = 0 .$$

Also $\gamma_o\vec{v}_1 \in (H^{1/2}(\Gamma))^n$. Therefore, we can apply Theorem 3.5 :

there exists $\vec{w}_1$ in $(H^1(\Omega))^n$ such that $\operatorname{div} \vec{w}_1 = 0$ and $\gamma_o\vec{w}_1 = \gamma_o\vec{v}_1$.

Then $\vec{v} = \vec{v}_1 - \vec{w}_1$ is the required function since $\vec{v} \in (H_o^1(\Omega))^n$ and $\operatorname{div} \vec{v} = q$.

Finally, it follows from the open mapping Theorem (cf. Yosida [46]) that the inverse of div is continuous from $L_o^2(\Omega)$ onto $\mathbf{V}^\perp$, therefore div is an isomorphism. ∎

Remark 3.5.

The usefulness of $L_o^2(\Omega)$ arises not only from this lemma, but also from the fact that $\forall\, q \in L_o^2(\Omega)$:

$$\|q\|_{o,\Omega} = \inf_{c \in \mathbb{R}} \|q+c\|_{o,\Omega} = \|\dot{q}\|_{L^2(\Omega)/\mathbb{R}} , \tag{3.24}$$

where $\dot{q}$ is the class of $L^2(\Omega)/\mathbb{R}$ containing q . As a consequence, it is often very handy to work with $L_o^2(\Omega)$ instead of $L^2(\Omega)/\mathbb{R}$. ∎

THEOREM 3.6.

Let $\vec{\ell}$ belong to $(H^{-1}(\Omega))^n$ and satisfy

$$< \vec{\ell},\vec{v} > = 0 \qquad \forall\, \vec{v} \in \mathbf{V} . \tag{3.25}$$

Then, there exists exactly one function φ in $L_o^2(\Omega)$ such that :

$$(3.26) \qquad \langle \vec{\ell},\vec{v} \rangle = \int_\Omega \varphi \operatorname{div} \vec{v}\, dx = - \langle \overrightarrow{\operatorname{grad}}\, \varphi,\vec{v} \rangle \qquad \forall\, \vec{v} \in (H_o^1(\Omega))^n .$$

PROOF.

Consider the following problem :

Find $\vec{u}$ in $V^\perp$ satisfying

$$(3.27) \qquad (\operatorname{div} \vec{u},\operatorname{div} \vec{v}) = \langle \vec{\ell},\vec{v} \rangle \qquad \forall\, \vec{v} \in V^\perp .$$

As div is an isomorphism from $V^\perp$ onto $L_o^2(\Omega)$, it follows that

$$\|\operatorname{div} \vec{v}\|_{o,\Omega}^2 \geq \alpha |\vec{v}|_{1,\Omega}^2 \qquad \forall\, \vec{v} \in V^\perp \text{, with } \alpha > 0 .$$

Hence, by the Lax-Milgram's Theorem, (3.27) has a unique solution $\vec{u}$ in $V^\perp$.

Then, hypothesis (3.25) implies that $\vec{u}$ also satisfies

$$(\operatorname{div} \vec{u},\operatorname{div} \vec{v}) = \langle \vec{\ell},\vec{v} \rangle \qquad \forall\, \vec{v} \in (H_o^1(\Omega))^n .$$

We set $\varphi = \operatorname{div} \vec{u} \in L_o^2(\Omega)$ and we find (3.26).

It remains to prove that φ is unique in $L_o^2(\Omega)$. But clearly, if $\varphi \in L_o^2(\Omega)$ and $(\varphi,\operatorname{div} \vec{v}) = 0 \quad \forall\, \vec{v} \in (H_o^1(\Omega))^n$, then $\varphi = 0$ since div maps $(H_o^1(\Omega))^n$ *onto* $L_o^2(\Omega)$. ■

The next theorem states another application of Lemma 3.2.

THEOREM 3.7.

There exists a constant $c > 0$ such that :

$$(3.28) \qquad \sup_{\vec{v} \in (H_o^1(\Omega))^n} \frac{(\varphi,\operatorname{div} \vec{v})}{|\vec{v}|_{1,\Omega}} \geq c \|\varphi\|_{o,\Omega} \qquad \forall\, \varphi \in L_o^2(\Omega) .$$

PROOF.

Let $\varphi \in L_o^2(\Omega)$. By virtue of Lemma 3.2, there exists a unique function $\vec{v} \in V^\perp$ such that $\varphi = \operatorname{div} \vec{v}$ and $|\vec{v}|_{1,\Omega} \leq c\, \|\varphi\|_{o,\Omega}$.

Hence

$$\frac{(\varphi,\operatorname{div} \vec{v})}{|\vec{v}|_{1,\Omega}} = \frac{\|\varphi\|_{o,\Omega}^2}{|\vec{v}|_{1,\Omega}} \geq \frac{1}{c} \|\varphi\|_{o,\Omega} . \qquad ■$$

We are now in a position to characterize $\mathbf{V}$ and $\mathbf{V}^{\perp}$.

DEFINITION 3.1.

Let $(-\Delta)^{-1}$ denote Green's operator related to Dirichlet's homogeneous problem for $-\Delta$ in $\mathbb{R}^n$, i.e. if $\vec{f} \in (H^{-1}(\Omega))^n$, then $(-\Delta)^{-1}\vec{f}$ is defined as *the* solution $\vec{u}$ of the problem

$$\vec{u} \in (H_0^1(\Omega))^n \ , \ -\Delta\vec{u} = \vec{f} \ \text{ in } \ \Omega \ .$$

THEOREM 3.8.

The space $\mathcal{V}$ is dense in $\mathbf{V}$ and

$$(3.29) \quad \mathbf{V} = \begin{cases} \{\vec{v} = \vec{\text{curl}}\,\varphi \ ; \ \varphi \in H^1(\Omega) \ \text{ with } \ \vec{\text{curl}}\,\varphi \in (H_0^1(\Omega))^2\} \ \text{ if } \ n = 2 \\ \{\vec{v} = \vec{\text{curl}}\,\vec{\varphi} \ ; \ \vec{\varphi} \in (H^1(\Omega))^3 \ \text{ with } \ \vec{\text{curl}}\,\vec{\varphi} \in (H_0^1(\Omega))^3\} \ \text{ if } \ n = 3, \end{cases}$$

$$(3.30) \qquad \mathbf{V}^{\perp} = \{\vec{v} = (-\Delta)^{-1}\overrightarrow{\text{grad}}\, q \ ; \ q \in L^2(\Omega)\} \ .$$

PROOF.

The characterization (3.29) of $\mathbf{V}$ follows immediately from Theorems 3.1. or 3.2. Let us check (3.30) ; let $\vec{u} \in \mathbf{V}^{\perp}$ and consider the mapping $\ell : \vec{v} \longmapsto (\overrightarrow{\text{grad}}\,\vec{u}, \overrightarrow{\text{grad}}\,\vec{v})$ for $\vec{v}$ in $(H_0^1(\Omega))^n$. Then ℓ is a continuous linear functional on $(H_0^1(\Omega))^n$ that vanishes on $\mathbf{V}$. According to Theorem 3.6, there exists q in $L_0^2(\Omega)$ such that

$$< \ell, \vec{v} > = (\overrightarrow{\text{grad}}\,\vec{u}, \overrightarrow{\text{grad}}\,\vec{v}) = -\,(q, \text{div}\,\vec{v}) \quad \forall\, \vec{v} \in (H_0^1(\Omega))^n \ .$$

Therefore,

$$-< \Delta\vec{u}, \vec{v} > = < \overrightarrow{\text{grad}}\, q, \vec{v} > \quad \forall\, \vec{v} \in (H_0^1(\Omega))^n \ .$$

Hence $\vec{u}$ satisfies $\vec{u} \in (H_0^1(\Omega))^n$ and $-\Delta\vec{u} = \overrightarrow{\text{grad}}\, q$ in Ω, i.e.

$$\vec{u} = (-\Delta)^{-1}\ \overrightarrow{\text{grad}}\, q \ .$$

Conversely, it is clear that $(-\Delta)^{-1}\overrightarrow{\text{grad}}\, q \in \mathbf{V}^{\perp}$ for every q in $L^2(\Omega)$.

Finally, the proof of the density of $\mathcal{V}$ in $\mathbf{V}$ is similar to that of the density of $\mathcal{V}$ in $\mathbf{H}$. ∎

Remark 3.6.

Since $\mathbf{V}$ is a closed subspace of $(H_o^1(\Omega))^n$, we have

$$(H_o^1(\Omega))^n = \mathbf{V} \oplus \mathbf{V}^\perp .$$

Hence, Theorem 3.8 implies that every function of $(H_o^1(\Omega))^n$ can be written in the form

$$\vec{v} = (-\Delta)^{-1}\,\overrightarrow{\text{grad}}\, q + \overrightarrow{\text{curl}}\, \vec{\varphi} . \tag{3.31}$$

The function q is uniquely determined in $L_o^2(\Omega)$ by (3.31). ∎

When $n = 2$, we can characterize entirely the stream function φ in (3.31) that satisfies $\varphi_{|\Gamma_o} = 0$. We introduce the following closed subspace of $H^2(\Omega)$:

$$\Phi = \{\chi \in H^2(\Omega)\ ;\ \chi_{|\Gamma_o} = 0\ ,\ \chi_{|\Gamma_i} \text{ is constant, } 1 \leqslant i \leqslant p, \text{ and } \frac{\partial \chi}{\partial \nu}_{|\Gamma} = 0\} .$$

Note that the semi-norm $\|\Delta\varphi\|_{o,\Omega}$ is a norm on Φ equivalent to $\|\varphi\|_{2,\Omega}$. Indeed, since $\overrightarrow{\text{grad}}\,\varphi \in [H_o^1(\Omega)]^2$ when $\varphi \in \Phi$, it can easily be shown like in section 1.5 and by virtue of Lemma 3.1 that

$$\|\Delta\varphi\|_{o,\Omega} = |\varphi|_{2,\Omega} \geqslant C_1\, |\varphi|_{1,\Omega} \geqslant C_2\, \|\varphi\|_{o,\Omega} .$$

Now, each function $\vec{v}$ of $[H_o^1(\Omega)]^2$ has exactly one stream function φ in Φ, and (3.31) implies that

$$\text{curl}\, \vec{v} = -\Delta\varphi + \text{curl}(-\Delta)^{-1}\,\overrightarrow{\text{grad}}\, q . \tag{3.32}$$

Let $\chi \in \Phi$ and let $\vec{w} = \overrightarrow{\text{curl}}\,\chi$, $\vec{w} \in \mathbf{V}$. From (3.32), we get:

$$(-\Delta\varphi, \text{curl}\,\vec{w}) = (\text{curl}\,\vec{v}, \text{curl}\,\vec{w}) - (\text{curl}(-\Delta)^{-1}\,\overrightarrow{\text{grad}}\, q, \text{curl}\,\vec{w}) .$$

But $(\text{curl}(-\Delta)^{-1}\,\overrightarrow{\text{grad}}\, q, \text{curl}\,\vec{w}) = \langle (-\Delta)^{-1}\,\overrightarrow{\text{grad}}\, q, \overrightarrow{\text{curl}}\ \text{curl}\,\vec{w} \rangle_{H_o^1, H^{-1}}$,

and

$$\overrightarrow{\text{curl}}\ \text{curl}\,\vec{w} = -\Delta\vec{w} + \overrightarrow{\text{grad}}\ \text{div}\,\vec{w} = -\Delta\vec{w} ,$$

since $\vec{w} \in \mathbf{V}$. Therefore

$$(\text{curl}(-\Delta)^{-1}\,\overrightarrow{\text{grad}}\, q, \text{curl}\,\vec{w}) = \langle (-\Delta)^{-1}\,\overrightarrow{\text{grad}}\, q, -\Delta\vec{w} \rangle = \langle \overrightarrow{\text{grad}}\, q, \vec{w} \rangle = 0$$

since $\vec{w} \in \mathbf{V}$. Hence, we have:

(3.33) $$(\Delta\varphi,\Delta\chi) = -(\text{curl } \vec{v},\Delta\chi) \qquad \forall\, \chi \in \Phi .$$

As $\|\Delta\varphi\|_{o,\Omega}$ is a norm on Φ, problem (3.33) has a unique solution φ in Φ and therefore this problem characterizes φ.

Let us interpret problem (3.33) in terms of a boundary value problem. By restricting χ to $\mathcal{D}(\Omega)$, we obtain :

(3.34) $$\Delta^2\varphi = -\Delta(\text{curl } \vec{v}) \quad \text{in} \quad H^{-2}(\Omega) .$$

Then, by taking the scalar product of both sides of (3.34) with χ in Φ, integrating by parts, and comparing with (3.33), we find (formally) :

$$< \Delta\varphi + \text{curl } \vec{v} , \frac{\partial\chi}{\partial\nu} >_\Gamma = < \frac{\partial}{\partial\nu}(\text{curl } \vec{v} + \Delta\varphi),\chi >_\Gamma \quad \forall\, \chi \in \Phi .$$

As $\chi \in \Phi$, this implies that, formally:

$$\int_{\Gamma_i} \frac{\partial}{\partial\nu}(\Delta\varphi + \text{curl } \vec{v})\, d\sigma = 0 \quad \text{for } 1 \leq i \leq p .$$

Thus, we have proved the following result :

COROLLARY 3.2.

1) Each function $\vec{v}$ in $[H_o^1(\Omega)]^2$ has exactly one stream function φ that vanishes on Γ_o and this function is the unique solution of the problem:

(3.33) $$\left\{ \begin{array}{l} \text{Find } \varphi \text{ in } \Phi \text{ such that} \\ (\Delta\varphi,\Delta\chi) = -(\text{curl } \vec{v},\Delta\chi) , \quad \forall\, \chi \in \Phi . \end{array} \right.$$

2) This stream function can be characterized equivalently as the solution of the boundary value problem :

$$\Delta^2\varphi = -\Delta(\text{curl } \vec{v}) \quad \text{in } \Omega ,$$

$$\varphi|_{\Gamma_o} = 0 \ , \quad \varphi|_{\Gamma_i} = \text{a constant } c_i \ , \quad 1 \leq i \leq p ,$$

$$\frac{\partial\varphi}{\partial\nu}\Big|_\Gamma = 0 \ , \quad \int_{\Gamma_i} \frac{\partial}{\partial\nu}(\Delta\varphi + \text{curl } \vec{v})\, d\sigma = 0 \ \text{ for } \ 1 \leq i \leq p ,$$

this last equation being formal.

§ 4 - ANALYSIS OF AN ABSTRACT VARIATIONAL PROBLEM

In this paragraph, we construct an abstract framework well adapted to the solution of a variety of linear boundary value problems with a constraint, like the Stokes problem. Two algorithms are proposed to deal with the constraint. Although they are introduced in connection with the continuous problem, they will prove to be useful mainly for solving the discretized problems.

4.1. Statement and solution of the problem.

Let X and M denote two real Hilbert spaces with the norms $\|\cdot\|_X$ and $\|\cdot\|_M$ respectively. Let X' and M' be their corresponding dual spaces and let $\|\cdot\|_{X'}$ and $\|\cdot\|_{M'}$ denote their dual norms. As usual, we denote the duality between X and X', or M and M', by $<\cdot,\cdot>$.

We introduce two *continuous bilinear* forms :

$$a(\cdot,\cdot) : X \times X \longmapsto \mathbb{R} \ , \ b(\cdot,\cdot) : X \times M \longmapsto \mathbb{R} \ ,$$

with norms

$$\|a\| = \sup_{\substack{u,v \in X \\ u\neq 0, v\neq 0}} \frac{a(u,v)}{\|u\|_X \|v\|_X} \ , \ \|b\| = \sup_{\substack{v\in X, \mu \in M \\ v\neq 0, \mu\neq 0}} \frac{b(v,\mu)}{\|v\|_X \|\mu\|_M}$$

Consider the following variational problem :

(Q) For ℓ given in X' and χ in M', find a pair (u,λ) in $X \times M$ such that :

$$\text{(4.1)} \quad a(u,v) + b(v,\lambda) = <\ell,v> \quad \forall v \in X$$

$$\text{(4.2)} \quad b(u,\mu) = <\chi,\mu> \quad \forall \mu \in M .$$

In order to study problem (Q), we require some extra notations. We associate with forms a and b two *continuous*, *linear* operators :

$$A \in \mathcal{L}(X \ ; X') \quad \text{and} \quad B \in \mathcal{L}(X \ ; M') \quad \text{defined by :}$$

$$\text{(4.3)} \qquad <Au,v> = a(u,v) \quad \forall u , v \in X ,$$

$$\text{(4.4)} \qquad <Bv,\mu> = b(v,\mu) \quad \forall v \in X , \ \forall \mu \in M .$$

Let $B' \in \mathcal{L}(M\ ;\ X')$ be the dual operator of B, i.e.

$$(4.5) \qquad < B'\mu, v > = < Bv, \mu > = b(v,\mu) \qquad \forall\, \mu \in M \ ,\ \forall\, v \in X \ .$$

It can be readily verified that

$$(4.6) \qquad \|A\|_{\mathcal{L}(X\ ;\ X')} = \|a\| \ ,\ \|B\|_{\mathcal{L}(X\ ;\ M')} = \|b\| .$$

With these operators, we have an equivalent formulation of problem (Q) :

$$(Q') \left\{ \begin{array}{l} \text{Find } (u,\lambda) \in X \times M \text{ satisfying :} \\ \qquad Au + B'\lambda = \ell \text{ in } X' , \\ \qquad Bu = \chi \text{ in } M' . \end{array} \right.$$

Next, we set $\mathbb{V} = \operatorname{Ker} B$ in X and more generally, for each $\chi \in M'$, we define the affine variety :

$$\mathbb{V}(\chi) = \{v \in X\ ;\ Bv = \chi\} .$$

Equivalently, we can write that

$$(4.7) \qquad \mathbb{V}(\chi) = \{v \in X\ ;\ b(v,\mu) = < \chi,\mu > \quad \forall\, \mu \in M\} \ ,$$

$$\mathbb{V} = \mathbb{V}(0) .$$

Moreover, the continuity of B implies that $\mathbb{V}$ is a closed subspace of X.

Now, with (Q) we associate the following problem :

$$(P) \left\{ \begin{array}{l} \text{Find } u \text{ in } \mathbb{V}(\chi) \text{ such that} \\ (4.8) \quad a(u,v) = < \ell,v > \quad \forall\, v \in \mathbb{V} . \end{array} \right.$$

Clearly, if $(u,\lambda) \in X \times M$ is a solution of (Q), then $u \in \mathbb{V}(\chi)$ and u is a solution of (4.8), i.e. u is a solution of (P). The rest of this section is devoted to show the converse of this statement and the existence and uniqueness of the solution, under suitable assumptions. For this, we define the polar set $\mathbb{V}^o$ of $\mathbb{V}$ by :

$$\mathbb{V}^o = \{g \in X'\ ;\ < g,v > = 0 \quad \forall\, v \in \mathbb{V}\} .$$

<u>LEMMA 4.1.</u>

<u>The three following properties are equivalent</u> :

(i) <u>there exists a constant</u> $\beta > 0$ <u>such that</u>

(4.9) $$\inf_{\mu \in M} \sup_{v \in X} \frac{b(v,\mu)}{\|v\|_X \|\mu\|_M} \geq \beta \ ;$$

(ii) <u>the operator</u> B' <u>is an isomorphism from</u> M <u>onto</u> V^o <u>and</u>

(4.10) $$\|B'\mu\|_{X'} \geq \beta \|\mu\|_M \quad \forall \mu \in M \ ;$$

(iii) <u>the operator</u> B <u>is an isomorphism from</u> $V^\perp$ <u>onto</u> M' <u>and</u>

(4.11) $$\|Bv\|_{M'} \geq \beta \|v\|_X \quad \forall v \in V^\perp .$$

<u>Proof</u>.

1) Let us show that (i) $\Leftrightarrow$ (ii).

By (4.5), statement (i) is equivalent to

$$\sup_{\substack{v \in X \\ v \neq 0}} \frac{\langle B'\mu, v \rangle}{\|v\|_X} \geq \beta \|\mu\|_M ,$$

that is, (4.9) is equivalent to (4.10). It remains to prove that B' is an isomorphism. Clearly, (4.10) implies that B' is a one-to-one operator from M onto its range $\mathcal{R}(B')$. Moreover, it also implies that the inverse of B' is continuous. Hence B' is an isomorphism from M onto $\mathcal{R}(B')$. Thus, we are led to prove that

$$\mathcal{R}(B') = V^o .$$

For this, we remark that $\mathcal{R}(B')$ is a closed subspace of X', since B' is an isomorphism. Therefore, we can apply the closed range theorem of Banach (cf. Yosida [46] p.205) which says that

$$\mathcal{R}(B') = (\text{Ker}(B))^o = V^o .$$

This proves part n°1.

2) (ii) $\Leftrightarrow$ (iii).

First, we observe that V^o can be identified isometrically with $(V^\perp)'$. Indeed, for $v \in X$ let $v^\perp$ denote the orthogonal projection of v on $V^\perp$. Then, with each $g \in (V^\perp)'$ we associate the element $\tilde{g}$ of X' defined by

$$\langle \tilde{g}, v \rangle = \langle g, v^\perp \rangle \qquad \forall v \in X .$$

Obviously $\tilde{g} \in V^o$ and it is easy to check that the correspondence $g \mapsto \tilde{g}$

maps isometrically $(V^{\perp})'$ onto V^{o}. This permits to identify $(V^{\perp})'$ and V^{o}. As a consequence, statements (ii) and (iii) are equivalent. ■

The condition (4.9) is usually called an " inf-sup " condition.

THEOREM 4.1.

Let us make the following hypotheses :

(i) There exists a constant $\alpha > 0$ such that

$$a(v,v) \geq \alpha \|v\|_X^2 \qquad \forall\, v \in V . \tag{4.12}$$

(ii) The bilinear form b satisfies the inf-sup condition (4.9).

Then problem (P) has a unique solution u in $V(\chi)$ and there exists a unique λ in M such that the pair (u,λ) is the unique solution of problem (Q). Moreover, the mapping $(\ell,\chi) \longmapsto (u,\lambda)$ is an isomorphism from $X' \times M'$ onto $X \times M$.

PROOF.

From (4.9) and Lemma 4.1, we see that there exists a unique element u_o in $V^{\perp}$ such that

$$Bu_o = \chi \quad \text{and} \quad \|u_o\|_X \leq \frac{1}{\beta}\|\chi\|_{M'} .$$

Therefore the following problem is equivalent to problem (P) :

$$(P') \quad \begin{cases} \text{Find } w = u - u_o \text{ in } V \text{ satisfying} \\ a(w,v) = \langle \ell, v \rangle - a(u_o,v) \quad \forall\, v \in V . \end{cases}$$

Since a is V-elliptic, we can apply the Lax-Milgram Theorem to (P'). Thus problem (P) has a unique solution u in $V(\chi)$ and

$$\|u\|_X \leq C_1(\|\ell\|_{X'} + \|\chi\|_{M'}) ,$$

where the constant C_1 depends only upon α, β and $\|a\|$.

Now, $\ell - Au$ belongs to V^{o} ; therefore, according to Lemma 4.1, there exists one and only one λ in M such that

$$B'\lambda = \ell - Au$$

and
$$\|\lambda\|_M \leq \frac{1}{\beta}\|\ell - Au\|_{X'} \leq C_2(\|\ell\|_{X'} + \|\chi\|_{M'}) .$$

Hence (u,λ) is the only solution of problem (Q).

The mapping $(\ell,\chi) \longmapsto (u,\lambda)$ is obviously an isomorphism from $X' \times M'$ onto $X \times M$. ∎

Remarks 4.1.

1) Under the hypotheses of Theorem 4.1, problems (P) and (Q) are equivalent.

2) If a is V-elliptic, then the inf - sup condition (4.9) is necessary as well as sufficient for the mapping $(u,\lambda) \longmapsto (\ell,\chi)$ to be an isomorphism from $X \times M$ onto $X' \times M'$. Indeed, we have already shown the sufficiency in the proof of Theorem 4.1 ; it remains to prove the necessity.

Let $\chi \in M'$ and let (u,λ) be the solution of (Q) with right-hand side $(0,\chi)$. Then $Bu = \chi$ and thus $\mathcal{R}(B) = M'$, so that B is a continuous and one-to-one mapping from $V^\perp$ onto M'. Therefore, B is an isomorphism from $V^\perp$ onto M'. Hence, by virtue of Lemma 4.1, the inf - sup condition is valid. ∎

4.2. A saddle-point approach

Under adequate hypotheses, it is possible to formulate problem (Q) in terms of a saddle-point problem.

In addition to the notations of the previous section, we introduce two quadratic functionals $J : X \longmapsto \mathbb{R}$ and $\mathcal{L} : X \times M \longmapsto \mathbb{R}$ defined by :

$$J(v) = \frac{1}{2}\, a(v,v) - \langle \ell,v \rangle \tag{4.13}$$

$$\mathcal{L}(v,\mu) = J(v) + b(v,\mu) - \langle \chi,\mu \rangle . \tag{4.14}$$

$\mathcal{L}$ is usually called the Lagrangian functional associated with problem (Q).

Consider the following problem :

(L) $\begin{cases} \text{Find a saddle-point } (u,\lambda) \text{ in } X \times M \text{ of the Lagrangian } \mathcal{L}, \\ \text{i.e. find a pair } (u,\lambda) \text{ in } X \times M \text{ such that :} \\ (4.15) \quad \mathcal{L}(u,\mu) \leqslant \mathcal{L}(u,\lambda) \leqslant \mathcal{L}(v,\lambda) \quad \forall v \in X ,\ \forall \mu \in M . \end{cases}$

THEOREM 4.2.

Under the hypotheses of Theorem 4.1 and if, moreover, the bilinear form a *is symmetric and semi positive definite on* X :

$$(4.16) \qquad a(v,v) \geq 0 \qquad \forall\, v \in X ,$$

then problem (L) has a unique solution (u,λ) *in* $X \times M$ *that is precisely the solution of* (Q).

Proof.

The first inequality in (4.15) can be written as follows :

$$b(u,\mu-\lambda) \leq \langle \chi,\mu-\lambda \rangle \qquad \forall\, \mu \in M .$$

As μ is any element of M this is equivalent to :

$$(4.2) \qquad b(u,\mu) = \langle \chi,\mu \rangle \qquad \forall\, \mu \in M .$$

Now, the second inequality in (4.15) is equivalent to

$$(4.17) \qquad \mathcal{L}(u,\lambda) = \inf_{v \in X} \mathcal{L}(v,\lambda) \quad .$$

Since, by hypothesis, a is symmetric, we have

$$\frac{\partial \mathcal{L}}{\partial v}(u,\lambda)\cdot v = a(u,v) + b(v,\lambda) - \langle \ell,v \rangle .$$

Futhermore, by (4.16) :

$$\frac{\partial^2 \mathcal{L}}{\partial v^2}(u,\lambda)(v,v) = a(v,v) \geq 0 .$$

Therefore $\mathcal{L}$ is a convex functional and its minimum (4.17) is characterized by the condition $\frac{\partial \mathcal{L}}{\partial v}(u,\lambda)\cdot v = 0$, i.e.

$$(4.1) \qquad a(u,v) + b(v,\lambda) = \langle \ell,v \rangle \qquad \forall\, v \in X .$$

Thus (u,λ) is a solution of (L) iff it is also a solution of (Q).
Hence the theorem is established. ∎

Remarks 4.2.

1) When the bilinear form a is symmetric and $\mathbb{V}$-elliptic, problem (P) may be viewed as an optimization problem. Indeed, the solution $u \in \mathbb{V}(\chi)$ of (P)

may be characterized as the unique element of $V(\chi)$ that satisfies :

$$J(u) = \inf_{v \in V(\chi)} J(v) .$$

Hence, λ appears to be *a* Lagrange multiplier associated with the constraint $u \in V(\chi)$.

2) The general optimization results yield the following equalities :

$$\mathcal{L}(u,\lambda) = \inf_{v \in X} \sup_{\mu \in M} \mathcal{L}(v,\mu) = \sup_{\mu \in M} \inf_{v \in X} \mathcal{L}(v,\mu). \qquad (4.18)$$ ■

4.3. Numerical solution by regularization.

We assume that hypotheses (4.12) and (4.9) hold. In addition to a and b we introduce a third continuous bilinear form $c(.,.) : M \times M \longmapsto \mathbb{R}$, M-elliptic, i.e. such that there exists a constant $\gamma > 0$ with :

$$c(\mu,\mu) \geq \gamma \|\mu\|_M^2 \qquad \forall \mu \in M. \qquad (4.19)$$

Let $C \in \mathcal{L}(M ; M')$ be defined by

$$< C\lambda,\mu > = c(\lambda, \mu) \qquad \forall \lambda , \mu \in M .$$

Let $\varepsilon > 0$ be a parameter which will tend to zero. We consider the problem :

$$(Q^\varepsilon) \begin{cases} \text{Find a pair } (u^\varepsilon,\lambda^\varepsilon) \in X \times M \text{ satisfying} \\ (4.20) \quad a(u^\varepsilon,v) + b(v,\lambda^\varepsilon) = < \ell,v > \qquad \forall v \in X \\ (4.21) \quad - \varepsilon c(\lambda^\varepsilon,\mu) + b(u^\varepsilon,\mu) = < \chi,\mu > \qquad \forall \mu \in M . \end{cases}$$

As C is non singular, equation (4.21) is equivalent to :

$$\lambda^\varepsilon = \frac{1}{\varepsilon} C^{-1}(Bu^\varepsilon-\chi) . \qquad (4.22)$$

Hence we can eliminate λ^ε from (4.20) and derive another problem, which is obviously equivalent to (Q^ε) :

$$(P^\varepsilon) \begin{cases} \text{Find } u^\varepsilon \in X \text{ such that} \\ (4.23) \quad a(u^\varepsilon,v) + \frac{1}{\varepsilon} < Bv,C^{-1}Bu^\varepsilon > = < \ell,v > + \frac{1}{\varepsilon} < Bv,C^{-1}\chi > \quad \forall v \in X . \end{cases}$$

Remark 4.3.

Clearly, when the bilinear forms are symmetric, solving problem (P^ε) is equivalent to finding $u^\varepsilon \in X$ such that

$$J_\varepsilon(u^\varepsilon) = \inf_{v \in X} J_\varepsilon(v) \quad \text{where} \quad J_\varepsilon(v) = J(v) + \frac{1}{2\varepsilon} < Bv-\chi, C^{-1}(Bv-\chi) > .$$

The expression $\frac{1}{2\varepsilon} < Bv - \chi, C^{-1}(Bv-\chi) >$ is a penalty term corresponding to the constraint $b(v,\mu) = 0$. Thus problem (P^ε) is a penalized version of problem (P). ■

THEOREM 4.3.

Under the hypotheses (4.9),(4.19) and if there exists a constant $\alpha > 0$ such that :

$$(4.24) \qquad a(v,v) + < Bv, C^{-1}Bv > \geq \alpha \|v\|_X^2 \qquad \forall\, v \in X ,$$

then problems (Q) and (Q^ε) both have one and only one solution . Moreover, the following error bound holds for every sufficiently small ε :

$$(4.25) \qquad \|u-u^\varepsilon\|_X + \|\lambda-\lambda^\varepsilon\|_M \leq K \ \varepsilon(\|\ell\|_{X'} + \|\chi\|_{M'}) ,$$

where the constant K depends upon α, β , $\|a\|$, $\|b\|$ and $\|c\|$ only .

PROOF.

Hypothesis (4.24) implies that a is V-elliptic. Hence problem (Q) has a unique solution (u,λ) in $X \times M$.

Now, it follows from (4.19) and (4.24) that problem (P^ε) has exactly one solution u^ε in X . Therefore, if we define λ^ε by (4.22) then $(u^\varepsilon,\lambda^\varepsilon)$ is the only solution of (Q^ε).

It remains to establish (4.25). From (4.20) and (4.1), (4.21) and (4.2), we get :

$$(4.26) \quad \left\{ \begin{array}{ll} a(u-u^\varepsilon,v) + b(v,\lambda-\lambda^\varepsilon) = 0 & \forall\, v \in X \\ b(u-u^\varepsilon,\mu) + \varepsilon\, c(\lambda^\varepsilon,\mu) = 0 & \forall\, \mu \in M . \end{array} \right.$$

The first equation, together with (4.9), yields :

$$\beta \| \lambda-\lambda^{\varepsilon} \|_M \leq \sup_{v \in X} \frac{b(v,\lambda-\lambda^{\varepsilon})}{\| v \|_X} \leq \| a \| \, \| u-u^{\varepsilon} \|_X ,$$

whence

$$\| \lambda-\lambda^{\varepsilon} \|_M \leq \frac{1}{\beta} \| a \| \, \| u-u^{\varepsilon} \|_X . \tag{4.27}$$

By taking $v = u - u^{\varepsilon}$ and $\mu = \lambda - \lambda^{\varepsilon}$ in (4.26), we find :

$$a(u-u^{\varepsilon},u-u^{\varepsilon}) = \varepsilon\, c(\lambda,\lambda-\lambda^{\varepsilon}) - \varepsilon c(\lambda-\lambda^{\varepsilon},\lambda-\lambda^{\varepsilon}) ,$$

$$\leq \varepsilon\, c(\lambda,\lambda-\lambda^{\varepsilon}) ,$$

owing to (4.19). Then (4.27) gives :

$$a(u-u^{\varepsilon},u-u^{\varepsilon}) \leq \varepsilon \| c \| \frac{\| a \|}{\beta} \| \lambda \|_M \| u-u^{\varepsilon} \|_X .$$

Besides that,

$$B(u-u^{\varepsilon}) = \chi - Bu^{\varepsilon} = - \varepsilon\, C\lambda^{\varepsilon} .$$

Therefore

$$< B(u-u^{\varepsilon}),C^{-1}B(u-u^{\varepsilon}) > = \varepsilon^2 c(\lambda^{\varepsilon},\lambda^{\varepsilon})$$

$$\leq \varepsilon^2 C_1 (\| \lambda \|_M + \| u-u^{\varepsilon} \|_X)^2 ,$$

where $C_1 = \| c \| \sup(1, \frac{\| a \|^2}{\beta^2})$. Hence, hypothesis (4.24) yields an inequality of the form

$$\alpha x^2 \leq \varepsilon^2 C_1 (\| \lambda \|_M + x)^2 + \varepsilon\, C_2 \| \lambda \|_M\, x ,$$

with $\| u-u^{\varepsilon} \|_X$ represented by x . If ε is sufficiently small, this amounts to :

$$\| u-u^{\varepsilon} \|_X \leq \varepsilon\, C_3 \| \lambda \|_M .$$

This, together with (4.27), prove the bound (4.25). ∎

4.4. Numerical solution by duality

We keep the notations of the previous section. The method proposed here is similar to that of the last section in that it splits the computation of u and λ . However, this is achieved by an iterative procedure. This method is based on Uzawa's classical algorithm (cf. for instance Arrow, Hurwicz & Uzawa [2]), which consists in constructing a sequence of functions $(u_m,\lambda_m) \in X \times M$ for all m , such that

$$a(u_{m+1},v) + b(v,\lambda_m) = < \ell,v > \quad \forall\, v \in X ,$$

$$- c(\lambda_{m+1}-\lambda_m,\mu) + \rho_m b(u_{m+1},\mu) = \rho_m < \chi,\mu > \quad \forall\, \mu \in M ,$$

where the parameters $\rho_m > 0$ are arbitrary.

Obviously, if a is V-elliptic and c is M-elliptic, these equations determine uniquely and separately u_{m+1} and λ_{m+1}.

We shall study a variant of this method which leads to a slightly more general algorithm obtained by a procedure known as the " augmented Lagrangian " technique. The idea is that, because $Bu_m - \chi$ is supposed to tend to zero, the ellipticity of a can be strengthened by adding to it the term $r < Bv, C^{-1}(Bv-\chi) >$ for some parameter $r > 0$. This yields the following problem.

Find a sequence (u_m, λ_m) in $X \times M$ satisfying :

$$(Q_m)\begin{cases} (4.28) \quad a(u_{m+1}, v) + r < Bv, C^{-1}Bu_{m+1} > + b(v, \lambda_m) = < \ell, v > + r < Bv, C^{-1}\chi > \\ \qquad\qquad\qquad\qquad\qquad\qquad\qquad\qquad \forall v \in X, \\ (4.29) \quad -c(\lambda_{m+1}-\lambda_m, \mu) + \rho_m b(u_{m+1}, \mu) = \rho_m < \chi, \mu > \qquad \forall \mu \in M. \end{cases}$$

Here again, the computation of u_{m+1} is dissociated from that of λ_{m+1}.

Theorem 4.4.

We assume that the hypotheses (4.9), (4.16), (4.19) and (4.24) hold. In addition we assume that the bilinear form $c(.,.)$ is symmetric and that there exists a constant $\alpha(r) > 0$ such that

$$(4.30) \qquad a(v,v) + r < Bv, C^{-1}Bv > \geq \alpha(r) \| Bv \|^2_{M'} \qquad \forall v \in X.$$

Then the algorithm (Q_m) uniquely determines a sequence (u_m, λ_m) in $X \times M$. Moreover, under the conditions :

$$(4.31) \qquad 0 < \inf_m \rho_m \leq \sup_m \rho_m < 2\gamma\alpha(r),$$

we have

$$\lim_{m \to \infty} \{\| u-u_m \|_X + \| \lambda-\lambda_m \|_M\} = 0.$$

PROOF.

From (4.19), (4.24) and (4.16), we derive readily that

$$a(v,v) + r < Bv, C^{-1}Bv > \geq \alpha \min(1,r) \| v \|^2_X \qquad \forall v \in X.$$

As a consequence, (4.28) defines a unique $u_{m+1} \in X$. Similarly, by virtue of (4.19), (4.29) defines λ_{m+1} uniquely in M.

Let us study the convergence of (Q_m). We set

$$v_m = u_m - u \ , \ \mu_m = \lambda_m - \lambda \ .$$

Then by subtracting (4.1) from (4.28) and (4.2) from (4.29), we get :

$$(4.32) \qquad a(v_{m+1},v) + r < Bv, C^{-1} Bv_{m+1} > = - b(v,\mu_m) \quad \forall\, v \in X \ ,$$

$$(4.33) \qquad c(\mu_{m+1} - \mu_m, \mu) = \rho_m b(v_{m+1},\mu) \quad \forall\, \mu \in M \ .$$

As c is symmetric, it satisfies the identity

$$c(\mu_{m+1} - \mu_m) = c(\mu_m) - c(\mu_{m+1}) - 2c(\mu_m - \mu_{m+1}, \mu_{m+1}) \ ,$$

where $c(\mu)$ stands for $c(\mu,\mu)$. With (4.33), this gives :

$$(4.34) \qquad c(\mu_{m+1}) - c(\mu_m) + c(\mu_{m+1} - \mu_m) = 2\, \rho_m b(v_{m+1}, \mu_{m+1}) \ .$$

From (4.32) and (4.34) we infer :

$$c(\mu_{m+1}) - c(\mu_m) + c(\mu_{m+1} - \mu_m) + 2\,\rho_m\, a(v_{m+1}, v_{m+1}) + 2r\rho_m < Bv_{m+1}, C^{-1} Bv_{m+1} >$$

$$= 2\rho_m b(v_{m+1}, \mu_{m+1} - \mu_m) \ .$$

Then hypothesis (4.19) and the fact that $\alpha(r) > 0$ yield :

$$c(\mu_{m+1}) - c(\mu_m) + \gamma \| \mu_{m+1} - \mu_m \|_M^2 + 2\rho_m \alpha(r) \| Bv_{m+1} \|_{M'}^2$$

$$\leq 2\rho_m \| Bv_{m+1} \|_{M'} \| \mu_{m+1} - \mu_m \|_M \ .$$

With the inequality $2ab \leq \gamma a^2 + \frac{1}{\gamma} b^2 \quad \forall\, a$ and $b > 0$, this becomes :

$$(4.35) \qquad c(\mu_{m+1}) - c(\mu_m) + \rho_m (2\alpha(r) - \frac{1}{\gamma}\rho_m) \| Bv_{m+1} \|_{M'}^2 \leq 0 \ .$$

If we choose ρ_m within the bounds (4.31) then there exists $\delta > 0$ such that

$$\rho_m (2\alpha(r) - \frac{1}{\gamma}\rho_m) \geq \delta \quad \forall\, m.$$

With this choice and by virtue of (4.11), (4.35) yields :

$$c(\mu_{m+1}) - c(\mu_m) + \delta\beta^2 \| v_{m+1} \|_X^2 \leq 0 \ .$$

Hence the sequence $c(\mu_m)$ is monotonically decreasing and bounded below by zero ; therefore it converges and

$$\lim_{m \to \infty} \| v_{m+1} \|_X^2 \leq \frac{1}{\delta \beta^2} \lim_{m \to \infty} (c(\mu_m) - c(\mu_{m+1})) = 0 \ .$$

It remains to prove that μ_m tends to zero. By applying (4.9) to the right-hand side of (4.32), we obtain :

$$\| \mu_m \|_M \leq \frac{1}{\beta} (\| a \| + r \| B \| \| C^{-1} B \|) \| v_{m+1} \|_X \ .$$

Hence $\lim_{m \to \infty} \mu_m = 0$ in M . ■

§ 5 - THEORY OF THE STOKES PROBLEM

In this paragraph, we establish the existence and uniqueness of the solution of the Stokes system and we give two variational formulations that we shall use later on for approximation purposes.

5.1. The " velocity-pressure " formulation

The Navier-Stokes equations describing the n-dimensional motion of a viscous and incompressible fluid are as follows :

$$(5.1) \qquad \rho\left(\frac{\partial u_i}{\partial t} + \sum_{j=1}^{n} u_j \frac{\partial u_i}{\partial x_j}\right) - \sum_{j=1}^{n} \frac{\partial}{\partial x_j} \sigma_{ij} = \rho f_i \quad , \quad 1 \leq i \leq n \ ,$$

with the incompressibility condition

$$(5.2) \qquad \operatorname{div} \vec{u} = \sum_{i=1}^{n} D_{ii}(\vec{u}) = 0 \quad ,$$

where

$$(5.3) \qquad \left\{ \begin{aligned} \sigma_{ij} &= - P \delta_i^j + 2\mu \, D_{ij}(\vec{u}) \\ D_{ij}(\vec{u}) &= \frac{1}{2} \left(\frac{\partial u_i}{\partial x_j} + \frac{\partial u_j}{\partial x_i}\right) \end{aligned} \right\} \quad 1 \leq i \, , \, j \leq n \ .$$

In these equations, the vector $\vec{u} = (u_1 , \ldots , u_n)$ is the velocity of the fluid, ρ is its density (assumed to be constant), $\mu > 0$ is its viscosity (also assumed to be constant) and P is its pressure ; (σ_{ij}) is the stress tensor and the vector $\vec{f} = (f_1 , \ldots , f_n)$ represents a density of body forces per unit mass (gravity, for instance).

We set

$$p = \frac{P}{\rho} \quad \text{and} \quad \nu = \frac{\mu}{\rho} \ . \tag{5.4}$$

Here, p is the kinematic pressure and ν the kinematic viscosity, but for the sake of simplicity they will be called in the sequel pressure and viscosity.

For the time being, we introduce two simplifications in equations (5.1). We only consider the steady state case, that is $\frac{\partial \vec{u}}{\partial t} = 0$, and furthermore, we assume that the velocity $\vec{u}$ is sufficiently small for ignoring the non-linear convection terms $u_j \frac{\partial u_i}{\partial x_j}$. Thus, we are led to the *Stokes system* of equations :

$$\left\{ \begin{aligned} & -2\nu \sum_{j=1}^{n} \frac{\partial}{\partial x_j} D_{ij}(\vec{u}) + \frac{\partial p}{\partial x_i} = f_i \ , \quad 1 \leqslant i \leqslant n \ , \\ & \sum_{i=1}^{n} D_{ii}(\vec{u}) = 0 \ . \end{aligned} \right. \tag{5.5}$$

Note that, when $\operatorname{div} \vec{u} = 0$, the following identity holds :

$$\sum_{j=1}^{n} \frac{\partial}{\partial x_j} D_{ij}(\vec{u}) = \frac{1}{2} \sum_{j=1}^{n} \left(\frac{\partial^2 u_i}{\partial x_j^2} + \frac{\partial^2 u_j}{\partial x_i \partial x_j} \right) = \frac{1}{2} \Delta u_i \ , \tag{5.6}$$

so that (5.5) can be written more conveniently

$$\left\{ \begin{aligned} & -\nu \Delta \vec{u} + \overrightarrow{\operatorname{grad}}\, p = \vec{f} \\ & \operatorname{div} \vec{u} = 0 \ . \end{aligned} \right. \tag{5.7}$$

The Stokes equations are linear, but nevertheless they deserve special attention, because of the incompressibility condition $\operatorname{div} \vec{u} = 0$.

THEOREM 5.1.

Let Ω be a bounded and connected subset of $\mathbb{R}^n$ with a Lipschitz continuous boundary Γ. Let $\vec{f}$ and $\vec{g}$ be two given functions in $(H^{-1}(\Omega))^n$ and $(H^{1/2}(\Gamma))^n$ respectively, such that

$$\int_\Gamma \vec{g} \cdot \vec{\nu} \, d\sigma = 0 \ . \tag{5.8}$$

Then, there exists one and only one pair of functions $(\vec{u}, p)$ in $(H^1(\Omega))^n \times L_0^2(\Omega)$ such that

$$(5.9)\quad \begin{cases} -\nu\Delta\vec{u} + \overrightarrow{\text{grad}}\, p = \vec{f} & \text{in } (H^{-1}(\Omega))^n , \\ \text{div}\, \vec{u} = 0 & \text{in } \Omega , \\ \vec{u} = \vec{g} & \text{on } \Gamma . \end{cases}$$

PROOF.

By virtue of (5.8) and Theorem 3.5 , there exists a function $\vec{u}_o$ in $(H^1(\Omega))^n$ such that

$$\text{div}\, \vec{u}_o = 0 \ , \ \vec{u}_{o|\Gamma} = \vec{g} .$$

Now, let us put problem (5.9) into the framework of paragraph 4 . We set

$$X = (H_o^1(\Omega))^n \ \text{ with } \|\cdot\|_X = |\cdot|_{1,\Omega} \ , \ M = L_o^2(\Omega) \ \text{ with } \|\cdot\|_M = \|\cdot\|_{o,\Omega} \ ,$$

$$a(\vec{u},\vec{v}) = 2\nu \left\{ \sum_{i,j=1}^{n} (D_{ij}(\vec{u}), D_{ij}(\vec{v})) \right\} ,$$

$$b(\vec{v},q) = -(q, \text{div}\, \vec{v}) ,$$

$$< \vec{\ell},\vec{v} > = < \vec{f},\vec{v} > - a(\vec{u}_o,\vec{v}) ,$$

$$\chi = 0 .$$

Then
$$\mathbb{V} = \{\vec{v} \in (H_o^1(\Omega))^n \ ; \ \text{div}\, \vec{v} = 0\} .$$

We must check that a is $\mathbb{V}$-elliptic and that b satisfies the inf-sup condition (4.9). First of all, since the operator D_{ij} is symmetric with respect to i and j , we have

$$(5.10)\quad a(\vec{u},\vec{v}) = 2\nu \sum_{i,j=1}^{n} (D_{ij}(\vec{u}), \frac{\partial v_i}{\partial x_j}) .$$

Next, an integration by parts shows that for $\vec{u}$ in X and $\vec{v}$ in $\mathbb{V}$, $a(\vec{u},\vec{v})$ can be written as

$$(5.11)\quad a(\vec{u},\vec{v}) = \nu \sum_{i,j=1}^{n} (\frac{\partial u_i}{\partial x_j} , \frac{\partial v_i}{\partial x_j}) = \nu(\overrightarrow{\text{grad}}\, \vec{u}, \overrightarrow{\text{grad}}\, \vec{v}) .$$

Thus
$$a(\vec{v},\vec{v}) = \nu|\vec{v}|_{1,\Omega}^2 ;$$

hence a is $\mathbb{V}$-elliptic. As far as b is concerned, the inf-sup condition says :

$$(5.12)\quad \sup_{\vec{v} \in (H_o^1(\Omega))^n} \frac{(q, \text{div}\, \vec{v})}{|\vec{v}|_{1,\Omega}} \geq \beta \|q\|_{o,\Omega} \qquad \forall\, q \in L_o^2(\Omega) .$$

This is precisely the conclusion of Theorem 3.7. Therefore, (5.12) is valid and we are in a position to apply Theorem 4.1 :

there exists one and only one pair of functions $(\vec{w},p)$ in $(H_o^1(\Omega))^n \times L_o^2(\Omega)$, such that

$$a(\vec{w},\vec{v}) + b(\vec{v},p) = \langle \vec{\ell},\vec{v} \rangle \quad \forall \vec{v} \in (H_o^1(\Omega))^n$$

and

$$b(\vec{w},q) = 0 \qquad \forall q \in L_o^2(\Omega) .$$

Then $(\vec{u} = \vec{w} + \vec{u}_o, p)$ is the solution of :

$$(5.13) \quad \begin{cases} \vec{u} - \vec{u}_o \in (H_o^1(\Omega))^n , \\ 2\nu \sum_{i,j=1}^{n} (D_{ij}(\vec{u}), D_{ij}(\vec{v})) - (p, \operatorname{div} \vec{v}) = \langle \vec{f},\vec{v} \rangle \quad \forall \vec{v} \in (H_o^1(\Omega))^n , \\ (q, \operatorname{div} \vec{u}) = (q, \operatorname{div} \vec{u}_o) = 0 \quad \forall q \in L_o^2(\Omega) . \end{cases}$$

Owing to Lemma 3.2 and the choice of $\vec{u}_o$, this last line implies that $\operatorname{div} \vec{u}_o = 0$. Hence there exists a unique pair $(\vec{u},p)$ in $(H^1(\Omega))^n \times L_o^2(\Omega)$ satisfying :

$$\vec{u}_{|\Gamma} = \vec{g} \;, \operatorname{div} \vec{u} = 0 \text{ and } (5.13).$$

It remains to show that this last problem is equivalent to (5.9). This is an immediate consequence of (5.10) and (5.6). ■

Remarks 5.1.

1) When $\vec{g} = 0$, the Stokes problem has the following (P) and (Q) formulations :

$$(Q) \quad \begin{cases} \text{Find a pair } (\vec{u},p) \in \mathbf{V} \times L_o^2(\Omega) \text{ satisfying :} \\ a(\vec{u},\vec{v}) - (p, \operatorname{div} \vec{v}) = \langle \vec{f},\vec{v} \rangle \quad \forall \vec{v} \in (H_o^1(\Omega))^n . \end{cases}$$

$$(P) \quad \begin{cases} \text{Find } \vec{u} \in \mathbf{V} \text{ such that :} \\ a(\vec{u},\vec{v}) = \langle \vec{f},\vec{v} \rangle \quad \forall \vec{v} \in \mathbf{V} , \end{cases}$$

where

$$a(\vec{u},\vec{v}) = 2\nu \left\{ \sum_{i,j=1}^{n} (D_{ij}(\vec{u}), D_{ij}(\vec{v})) \right\}$$

or equivalently

$$a(\vec{u},\vec{v}) = \nu(\overrightarrow{\operatorname{grad}}\, \vec{u}, \overrightarrow{\operatorname{grad}}\, \vec{v}) \quad \forall \vec{u} \in \mathbf{V} , \forall \vec{v} \in (H_o^1(\Omega))^n .$$

2) Of course, the choice $M = L_o^2(\Omega)$ is only a matter of convenience, and we can just as well take $M = L^2(\Omega)/\mathbb{R}$ (cf. remark 3.5). ■

The next theorem concerns the regularity of the solution of the Stokes problem when the boundary is sufficiently smooth. Part 1 is proved in Temam [44] and part 2 in Grisvard [28].

THEOREM 5.2.

1) _In addition to the hypotheses of Theorem 5.1, suppose that_ Γ _is of class_ $\mathscr{C}^2$, $\vec{g} = 0$ _and_ $\vec{f}$ _is given in_ $(L^r(\Omega))^n$ _for_ $1 < r \leq 2$. _Then the Stokes problem (5.9) has a unique solution_ $(\vec{u},p)$ _in_ $(W^{2,r}(\Omega))^n \times (W^{1,r}(\Omega) \cap L_o^2(\Omega))$ _and there exists a constant_ C_r _independent of_ $\vec{u}$, p _and_ $\vec{f}$ _such that_ :

$$\|\vec{u}\|_{2,r,\Omega} + \|p\|_{1,r,\Omega} \leq C_r \|\vec{f}\|_{o,r,\Omega}. \tag{5.14}$$

2) _When_ Γ _is only Lipschitz continuous, this conclusion is still valid provided_ $n = 2$ _and_ Ω _is convex_.

The Stokes problem (5.9) can also be expressed as a saddle-point problem. With the above notations, we set

$$J(\vec{v}) = \frac{1}{2} a(\vec{v},\vec{v}) - < \vec{f},\vec{v} > , \tag{5.15}$$

$$\mathcal{L}(\vec{v},q) = J(\vec{v}) - (q, \operatorname{div} \vec{v}) . \tag{5.16}$$

As a is symmetric and $(H_o^1(\Omega))^n$ - elliptic, we have the following result.

THEOREM 5.3.

Under the hypotheses of Theorem 5.1, the solution $(\vec{u},p)$ _of_ (5.9) _is characterized by_ :

$$\begin{cases} \mathcal{L}(\vec{u},p) = \inf\limits_{\vec{v} \in (H^1(\Omega))^n, \gamma_o\vec{v} = \vec{g}} \left\{ \sup\limits_{q \in L_o^2(\Omega)} \mathcal{L}(\vec{v},q) \right\} \\ \quad = \sup\limits_{q \in L_o^2(\Omega)} \left\{ \inf\limits_{\vec{v} \in (H^1(\Omega))^n, \gamma_o\vec{v} = \vec{g}} \mathcal{L}(\vec{v},q) \right\}. \end{cases} \tag{5.17}$$

Furthermore, $\vec{u}$ _is characterized by_ :

$$J(\vec{u}) = \inf_{\vec{v} \in (H^1(\Omega))^n, \ \gamma_o\vec{v} = \vec{g}, \operatorname{div} \vec{v} = 0} J(\vec{v}) . \tag{5.18}$$

<u>PROOF</u>. Adapting Theorem 4.2 to the above situation, we find that $(\vec{u}-\vec{u}_o,p)$ is the saddle point of the Lagrangian functional :

$$\mathcal{L}_o(\vec{v},q) = \frac{1}{2}\, a(\vec{v},\vec{v}) - \langle \vec{f},\vec{v} \rangle + a(\vec{u}_o,\vec{v}) - (q,\mathrm{div}\,\vec{v})$$

over $(H^1_o(\Omega))^n \times L^2_o(\Omega)$. By writing that $(\vec{u}-\vec{u}_o,p)$ is the saddle point of $\mathcal{L}_o$ (inequalities (4.15)), and expanding, we get, on account of $\mathrm{div}\,\vec{u}_o = 0$:

$$\frac{1}{2}\, a(\vec{u},\vec{u}) - \langle \vec{f},\vec{u} \rangle - (q,\mathrm{div}\,\vec{u}) \leq \frac{1}{2}\, a(\vec{u},\vec{u}) - \langle \vec{f},\vec{u} \rangle - (p,\mathrm{div}\,\vec{u})$$

$$\leq \frac{1}{2}\, a(\vec{v}+\vec{u}_o,\vec{v}+\vec{u}_o) - \langle \vec{f},\vec{v}+\vec{u}_o \rangle - (p,\mathrm{div}(\vec{v}+\vec{u}_o)).$$

Hence $(\vec{u},p)$ is the saddle-point of the Lagrangian (5.16) over $(\vec{u}_o + (H^1_o(\Omega))^n) \times L^2_o(\Omega)$. Then, the desired result (5.17) is established by virtue of :

$$\vec{u}_o + (H^1_o(\Omega))^n = \{\vec{v} \in (H^1(\Omega))^n \ ;\ \vec{v}_{|\Gamma} = \vec{g}\},$$

and (4.18).

The proof of (5.18) is much the same. ∎

5.2. <u>The " stream function " formulation</u>

Here, we consider only the case $n = 2$. We keep the hypotheses of section 5.1 and moreover we assume that

$$\int_{\Gamma_i} \vec{g}.\vec{\nu}\, d\sigma = 0 \quad \text{for} \quad 0 \leq i \leq p\,, \tag{5.19}$$

where, as usual, Γ_i for $0 \leq i \leq p$ denote the components of Γ (cf. figure 1). Then, according to Theorem 3.1, the velocity vector $\vec{u}$ is the curl of a stream function ψ. We are going to show that the stream function can be characterized as the solution of a non-homogeneous biharmonic problem in Ω.

The stream function ψ is unique up to an additive constant. But, as $\psi \in H^2(\Omega) \subset \mathcal{C}^o(\bar{\Omega})$, ψ can be determined by fixing its value on one point of $\bar{\Omega}$. At first, we set $\psi(x_o) = 0$, where x_o is an arbitrary point of Γ_o. Next, we *choose* a function χ in $H^{3/2}(\Gamma)$ that satisfies :

$$\frac{\partial \chi}{\partial \tau} = \vec{g}\cdot\vec{\nu} \ \text{ on } \ \Gamma \ , \quad \chi(x_o) = 0\,. \tag{5.20}$$

Since $\frac{\partial\psi}{\partial\tau} = \vec{g}\cdot\vec{\nu}$ on Γ, it follows that ψ coincides with χ on Γ_o and differs from χ by a constant on the other components of Γ. More precisely :

$$(5.21) \quad \begin{cases} \psi = \chi \text{ on } \Gamma_o , \\ \psi = \chi + c_i \text{ on } \Gamma_i \text{ for } 1 \leqslant i \leqslant p , \end{cases}$$

where the c_i are fixed, unknown constants.

THEOREM 5.4.

Let $n = 2$ and let the hypotheses of Theorem 5.1 be satisfied. Then, under the condition

$$(5.19) \quad \int_{\Gamma_i} \vec{g}.\vec{\nu} \, d\sigma = 0 \quad \text{for } 0 \leqslant i \leqslant p ,$$

there exists a unique function $\psi \in H^2(\Omega)$ characterized by the equations :

$$(5.22) \quad \nu(\Delta\psi,\Delta\varphi) = \langle \vec{f}, \overrightarrow{\text{curl}}\, \varphi \rangle \quad \forall \varphi \in \Phi \quad \text{(cf. section 3.3)} ,$$

$$(5.21) \quad \psi = \chi \text{ on } \Gamma_o , \quad \psi = \chi + c_i \text{ on } \Gamma_i , \quad 1 \leqslant i \leqslant p ,$$

$$(5.23) \quad \frac{\partial\psi}{\partial\nu} = -\vec{g}.\vec{\tau} \text{ on } \Gamma ,$$

where χ is chosen according to (5.20).

PROOF.

From Theorems 4.1 and 5.1, we know that the first argument $\vec{u}$ of the solution of (5.9) is also the only solution of the problem :

$$(P_g) \quad \begin{cases} \text{Find } \vec{u} \in (H^1(\Omega))^2 \text{ such that} \\ a(\vec{u},\vec{v}) = \langle \vec{f},\vec{v} \rangle \quad \forall \vec{v} \in \mathbf{V} \\ \text{div}\, \vec{u} = 0 \quad , \quad \vec{u} = \vec{g} \text{ on } \Gamma . \end{cases}$$

Besides that, according to Corollary 3.2 , $\vec{v} \in \mathbf{V}$ iff there exists a unique stream function $\varphi \in \Phi$ such that $\vec{v} = \overrightarrow{\text{curl}}\, \varphi$. Let us express the form $a(\vec{u},\vec{v})$ in terms of stream functions. First, recall the following identities :

$$(5.24) \quad - \Delta\vec{w} = \overrightarrow{\text{curl}}(\text{curl}\, \vec{w}) \quad \forall \vec{w} \in (\mathcal{D}'(\Omega))^2 \text{ with div}\, \vec{w} = 0 ,$$

and

$$(5.25) \quad - \Delta\theta = \text{curl}(\overrightarrow{\text{curl}}\, \theta) \quad \forall \theta \in \mathcal{D}'(\Omega).$$

Next, let $\vec{v} \in \mathcal{V} = \{\vec{v} \in (\mathcal{D}(\Omega))^2 \; ; \; \text{div}\, \vec{v} = 0\}$. Then,

$$(\overrightarrow{\text{grad}}\, \vec{u}, \overrightarrow{\text{grad}}\, \vec{v}) = -(\vec{u}, \Delta\vec{v}) = (\vec{u}, \overrightarrow{\text{curl}}\, \text{curl}\, \vec{v}) = (\text{curl}\, \vec{u}, \text{curl}\, \vec{v}).$$

As $\mathcal{V}$ is dense in $\mathbf{V}$ according to Theorem 3.8, we get

$$(5.26) \qquad (\overrightarrow{\text{grad}}\, \vec{u}, \overrightarrow{\text{grad}}\, \vec{v}) = (\text{curl}\, \vec{u}, \text{curl}\, \vec{v}) \quad \forall\, \vec{u} \in (H^1(\Omega))^2 \; , \; \forall\, \vec{v} \in \mathbf{V} \, .$$

Finally, (5.11), (5.25) and (5.26) yield :

$$a(\vec{u}, \vec{v}) = \nu(\Delta\psi, \Delta\varphi) \; , \text{ where } \psi \text{ is the stream function of } \vec{u} \, .$$

Therefore the stream function ψ satisfies (5.22) $\forall\, \varphi \in \Phi$. Moreover (5.23) is a consequence of the boundary condition $\gamma_o \vec{u} = \vec{g}$. ■

It remains to interpret problem (5.21)(5.22)(5.23). By applying (formally) Green's formula, we can easily show that ψ is the only solution of the boundary value problem :

$$\nu\Delta^2\psi = \text{curl}\, \vec{f} \, ,$$

$$\psi = \chi \text{ on } \Gamma_o \; , \quad \psi = c_i + \chi \text{ on } \Gamma_i \text{ for } 1 \leqslant i \leqslant p \, ,$$

$$\frac{\partial\psi}{\partial\nu} = -\vec{g}\cdot\vec{\tau} \quad \text{on } \Gamma$$

and

$$\int_{\Gamma_i} \left(\nu \frac{\partial\Delta\psi}{\partial n} - \vec{f}\cdot\vec{\tau}\right) d\sigma = 0 \quad \text{for } 1 \leqslant i \leqslant p \, ,$$

where, in order to avoid confusion, $\frac{\partial}{\partial n}$ denotes here the normal derivative.

CHAPTER II

NUMERICAL SOLUTION OF THE STOKES PROBLEM
A CLASSICAL METHOD

§ 1. AN ABSTRACT APPROXIMATION RESULT

This short paragraph is devoted to the approximation of the abstract variational problem analysed in § 4, Chapter I. We keep here the same notation and we put the problem in exactly the same situation. In particular, we assume that the hypotheses (i) and (ii) of Theorem 4.1 are satisfied.

Let h denote a discretization parameter tending to zero and, for each h, let X_h and M_h be two finite-dimensional spaces such that

$$X_h \subset X, \quad M_h \subset M.$$

We approximate problem (Q) by :

$$(Q_h)\begin{cases} \text{Find a pair } (u_h,\lambda_h) \text{ in } X_h \times M_h \text{ satisfying} \\ (1.1) \qquad a(u_h,v_h) + b(v_h,\lambda_h) = \langle \ell, v_h \rangle \quad \forall v_h \in X_h, \\ (1.2) \qquad b(u_h,\mu_h) = \langle \chi, \mu_h \rangle \quad \forall \mu_h \in M_h. \end{cases}$$

For each $\chi \in M'$, we define the finite-dimensional analogue of $V(\chi)$:

$$(1.3) \qquad V_h(\chi) = \{v_h \in X_h \ ; \ b(v_h,\mu_h) = \langle \chi, \mu_h \rangle \quad \forall \mu_h \in M_h\}$$

and we set $V_h = V_h(0)$, i.e.

$$(1.4) \qquad V_h = \{v_h \in X_h \ ; \ b(v_h,\mu_h) = 0 \quad \forall \mu_h \in M_h\}.$$

Right away, we remark that since M_h is a proper subspace of M then, in general, $V_h \not\subset V$ and $V_h(\chi) \not\subset V(\chi)$.

Like in the continuous case, we associate with (Q_h) the following problem :

$$(P_h)\begin{cases} \text{Find } u_h \in V_h(\chi) \text{ such that} \\ (1.5) \qquad a(u_h,v_h) = \langle \ell, v_h \rangle \quad \forall v_h \in V_h. \end{cases}$$

As $V_h \not\subset V$, problem (P_h) may be viewed as an *external approximation of* (P). Here again, the first component u_h of any solution (u_h,λ_h) of problem (Q_h) is also a solution of (P_h). The converse is proved as part of the next theorem.

THEOREM 1.1.

1°/ Assume that the following conditions hold :

(i) $V_h(\chi)$ is not empty ;

(ii) there exists a constant $\alpha^\star > 0$ such that :

$$a(v_h,v_h) \geq \alpha^\star \|v_h\|_X^2 \quad \forall v_h \in V_h. \tag{1.6}$$

Then problem (P_h) has a unique solution $u_h \in V_h(\chi)$ and there exists a constant C_1 depending only upon $\alpha^\star$, $\|a\|$ and $\|b\|$ such that the "error bound" holds :

$$\|u-u_h\|_X \leq C_1\{ \inf_{v_h \in V_h(\chi)} \|u-v_h\|_X + \inf_{\mu_h \in M_h} \|\lambda-\mu_h\|_M\}. \tag{1.7}$$

2°/ Assume that hypothesis (ii) holds and, in addition, that :

(iii) there exists a constant $\beta^\star > 0$ such that

$$\sup_{v_h \in X_h} \frac{b(v_h,\mu_h)}{\|v_h\|_X} \geq \beta^\star \|\mu_h\|_M \quad \forall \mu_h \in M_h. \tag{1.8}$$

Then $V_h(\chi) \neq \phi$ and there exists a unique λ_h in M_h so that (u_h,λ_h) is the only solution of (Q_h). Furthermore, there exists a constant C_2 depending only upon $\alpha^\star$, $\beta^\star$, $\|a\|$ and $\|b\|$ such that

$$\|u-u_h\|_X + \|\lambda-\lambda_h\|_M \leq C_2\{ \inf_{v_h \in X_h} \|u-v_h\|_X + \inf_{\mu_h \in M_h} \|\lambda-\mu_h\|_M\}. \tag{1.9}$$

Proof

1°/ As $V_h(\chi)$ is not empty, we choose a u_h^0 in $V_h(\chi)$ and we solve the problem : find z_h in V_h such that

$$a(z_h,v_h) = \langle \ell, v_h \rangle - a(u_h^0,v_h) \quad \forall v_h \in V_h.$$

From (1.6), this problem has a unique solution z_h, and therefore, $u_h = z_h + u_h^0$ is the unique solution of problem (P_h).

Let w_h be an arbitrary element of $V_h(\chi)$; then $v_h = u_h - w_h \in V_h$ and

$$a(v_h,v_h) = \langle \ell, v_h \rangle - a(w_h,v_h). \tag{1.10}$$

As $v_h \in X_h$, we can take $v = v_h$ in equation (4.1), Chapter I and substitute in (1.10). This yields :

$$a(v_h, v_h) = a(u-w_h, v_h) + b(v_h, \lambda).$$

Moreover, since $v_h \in V_h$, we have $b(v_h, \mu_h) = 0 \quad \forall \mu_h \in M_h$. Hence

$$(1.11) \qquad a(v_h, v_h) = a(u-w_h, v_h) + b(v_h, \lambda-\mu_h) \quad \forall \mu_h \in M_h.$$

The ellipticity of a and the continuity of a and b yield :

$$\|v_h\|_X \leq \frac{1}{\alpha^*}\left(\|a\| \|u-w_h\|_X + \|b\| \|\lambda-\mu_h\|_M\right).$$

Therefore,

$$\|u-u_h\|_X \leq \left(1+\frac{\|a\|}{\alpha^*}\right)\|u-w_h\|_X + \frac{\|b\|}{\alpha^*}\|\lambda-\mu_h\|_M ,$$

$$\forall w_h \in V_h(\chi), \quad \forall \mu_h \in M_h.$$

This yields (1.7) with $C_1 = \sup\left(1+\frac{\|a\|}{\alpha^*}, \frac{\|b\|}{\alpha^*}\right)$.

2°/ Let us apply Lemma 4.1, Chapter I, to the particular case of X_h and M_h. Let $(\cdot,\cdot)_M$ denote the scalar product on M associated with $\|\cdot\|_M$ and let $B_h \in \mathcal{L}(X_h; M_h)$ be defined by $(B_h v_h, \mu_h)_M = b(v_h, \mu_h)$. Then hypothesis (iii) implies that B_h is an isomorphism from $V_h^\perp$ (taken in X_h) onto M_h. Therefore $V_h(\chi)$ is not empty and according to part 1, problem (P_h) has a unique solution u_h. Furthermore, it follows from Theorem 4.1 that there exists a unique λ_h in M_h such that (u_h, λ_h) is the only solution of (Q_h).

To derive the error bound (1.9), we shall first prove that

$$(1.12) \qquad \inf_{w_h \in V_h(\chi)} \|u-w_h\|_X \leq \left(1+\frac{\|b\|}{\beta^*}\right) \inf_{v_h \in X_h} \|u-v_h\|_X.$$

By virtue of Lemma 4.1 (iii) and (4.11), for each $v_h \in X_h$, there exists a unique z_h in $V_h^\perp$ satisfying

$$b(z_h, \mu_h) = b(u-v_h, \mu_h) \qquad \forall \; \mu_h \in M_h ,$$

$$\|z_h\|_X \leq \frac{\|b\|}{\beta^*} \|u-v_h\|_X .$$

Thus $\qquad w_h = v_h + z_h$ belongs to $V_h(\chi)$ and moreover,

$$\|u-w_h\|_X \leq \left(1+\frac{\|b\|}{\beta^*}\right)\|u-v_h\|_X .$$

As v_h is an arbitrary element of X_h , this implies (1.12).

It remains to evaluate $\|\lambda-\lambda_h\|_M$. From (1.1) and (4.1), Chapter I, we get

$$b(v_h,\lambda_h) = a(u-u_h,v_h) + b(v_h,\lambda) \quad \forall v_h \in X_h.$$

Therefore

$$b(v_h,\lambda_h-\mu_h) = a(u-u_h,v_h) + b(v_h,\lambda-\mu_h) \quad \forall v_h \in X_h, \quad \mu_h \in M_h.$$

Then (1.8) yields :

$$\|\lambda_h-\mu_h\|_M \leq \frac{1}{\beta^*} \sup_{v_h\in X_h} \frac{|a(u-u_h,v_h)+b(v_h,\lambda-\mu_h)|}{\|v_h\|_X}$$

$$\leq \frac{1}{\beta^*}\left(\|a\|\,\|u-u_h\|_X + \|b\|\,\|\lambda-\mu_h\|_M\right).$$

Hence

$$\|\lambda-\lambda_h\|_M \leq \frac{1}{\beta^*}\left[\|a\|\,\|u-u_h\|_X + \left(\beta^* + \|b\|\right) \inf_{\mu_h\in M_h} \|\lambda-\mu_h\|_M\right]. \tag{1.13}$$

Then bound (1.9) follows immediately from (1.7), (1.12) and (1.13). ∎

Remark 1.1.

Since usually $V_h \not\subset V$, the ellipticity of form a on V does not necessarily carry over to V_h. As a consequence, hypothesis (1.6) must be checked in each particular case (except of course when $V_h \subset V$).

As far as the inf-sup condition is concerned, it is clear that the continuous condition does not imply its discrete counterpart (1.8). In fact, (1.8) acts as a compatibility condition between spaces X_h and M_h. In practice, it turns out often that the condition (1.8) is not trivial to check. ∎

Remark 1.2.

It is possible to improve the bound (1.7) without making use of (1.8). Indeed, by applying (1.6) to (1.11) we get :

$$\|u_h-w_h\|_X \leq \frac{1}{\alpha^*}\left[\|a\|\,\|u-w_h\|_X + \sup_{v_h\in V_h} \frac{b(v_h,\lambda-\mu_h)}{\|v_h\|_X}\right].$$

Hence

$$\|u-u_h\|_X \leq C\left\{ \inf_{w_h\in V_h(\chi)} \|u-w_h\|_X + \inf_{\mu_h\in M_h} \sup_{v_h\in V_h} \frac{b(v_h,\lambda-\mu_h)}{\|v_h\|_X} \right\}, \tag{1.14}$$

where the constant C depends only upon $\alpha^\star$ and $\|a\|$. Note that the term

$$\inf_{\mu_h\in M_h} \sup_{v_h\in V_h} \frac{b(v_h,\lambda-\mu_h)}{\|v_h\|_X}$$

takes into account the error committed by using an external approximation. In particular, it vanishes when $V_h\subset V$. ∎

<u>Remark 1.3.</u>

If, besides hypotheses (1.6) and (1.8), we assume that the form a is symmetric and semi positive definite on X, then we can relate problems (P_h) and (Q_h) to optimization problems. As in the continuous case, and with the same notations, it can be shown that

$$J(u_h) = \inf_{v_h\in V_h(\chi)} J(v_h)$$

and

$$\mathcal{L}(u_h,\lambda_h) = \inf_{v_h\in X_h} \sup_{\mu_h\in M_h} \mathcal{L}(v_h,\mu_h) = \sup_{\mu_h\in M_h} \inf_{v_h\in X_h} \mathcal{L}(v_h,\mu_h).$$ ∎

Theorem 1.1 readily yields the following general convergence result.

<u>COROLLARY 1.1.</u>

1°/ <u>Assume that the following hypotheses hold</u> :

(i) <u>there exists a dense subvariety</u> $\mathcal{V}(\chi)$ <u>of</u> $V(\chi)$, <u>a dense subspace</u> $\mathcal{M}$ <u>of M and two mappings</u> $r_h:\mathcal{V}(\chi)\mapsto V_h(\chi)$ <u>and</u> $\rho_h:\mathcal{M}\mapsto M_h$ <u>such that</u>

$$(1.15)\qquad \begin{cases} \lim\limits_{h\to o} \|r_h v-v\|_X = 0 & \forall v\in \mathcal{V}(\chi) \\ \lim\limits_{h\to o} \|\rho_h\mu-\mu\|_M = 0 & \forall \mu\in \mathcal{M}; \end{cases}$$

(ii) <u>the form</u> a <u>satisfies (1.6) with a constant</u> $\alpha^\star$ <u>independent of</u> h.

Then

$$\lim_{h\to o} \|u-u_h\|_X = 0.$$

2°/ <u>Assume in addition that</u> :

(iii) <u>there exists a dense subspace</u> $\mathcal{X}$ <u>of</u> X <u>and a mapping still denoted by</u> $r_h:\mathcal{X}\mapsto X$ <u>such that</u>

(1.16) $$\lim_{h\to o} \| r_h v - v \|_X = 0 \quad \forall v \in \mathcal{X} ;$$

(iv) the form b satisfies (1.8) with a constant β^* independent of h.

Then

$$\lim_{h\to o} \{ \| u-u_h \|_X + \| \lambda-\lambda_h \|_M \} = 0.$$

Remark 1.4.

In this corollary, conditions (1.6) and (1.8) may be viewed as stability conditions, whereas conditions (1.15) and (1.16) are consistency conditions. ∎

We end this paragraph by extending the classical duality argument of Aubin [3] Nitsche [40] to the case of problems (P) and (P_h). For this, we introduce a Hilbert space H, with scalar product (.,.) and associated norm $|\cdot|$ such that

$X \subset H$ with continuous imbedding and X is dense in H.

We identify H with its dual space H' for the scalar product (.,.). Therefore, H can be identified with a subspace of X' :

$H \subset X'$ with continuous and dense imbedding.

The next theorem evaluates $|u-u_h|$.

THEOREM 1.2.

Suppose that the conditions (4.9) and (4.12) of Chapter I and (1.6) hold. Let (u,λ) be the solution of (Q), u_h the solution of (P_h) and for each g in H let $(\phi_g,\xi_g) \in X \times M$ be the solution of

(1.17) $$\begin{cases} a(v,\phi_g) + b(v,\xi_g) = (g,v) & \forall v \in X, \\ b(\phi_g,\mu) = 0 & \forall \mu \in M. \end{cases}$$

Then there exists a constant C, depending only upon $\|a\|$ and $\|b\|$, such that :

(1.18) $$|u-u_h| \leq C\{ \| u-u_h \|_X + \inf_{\mu_h \in M_h} \| \lambda-\mu_h \|_M \} \times \sup_{g\in H} \{ \frac{1}{|g|} (\inf_{\phi_h \in V_h} \| \phi_g-\phi_h \|_X + \inf_{\xi_h \in M_h} \| \xi_g-\xi_h \|_M)\} .$$

Proof

We have

$$|u-u_h| = \sup_{g\in H} \frac{|(g,u-u_h)|}{|g|} .$$

Now, it is clear that (1.17) has exactly one solution (ϕ_g,ξ_g) for each g in H. Therefore we can write :

$$(1.19) \qquad (g,u-u_h) = a(u-u_h,\phi_g) + b(u-u_h,\xi_g).$$

But, as u_h is the solution of (P_h) and (u,λ) that of (Q), we have for each ϕ_h in V_h and μ_h in M_h :

$$a(u-u_h,\phi_h) = a(u,\phi_h) - \langle \ell,\phi_h\rangle = -b(\phi_h,\lambda) = -b(\phi_h,\lambda-\mu_h).$$

Besides that,

$$b(\phi_g,\lambda-\mu_h) = 0,$$

since $\phi_g \in V$; and for every ξ_h in M_h, we have

$$b(u-u_h,\xi_h) = 0,$$

since $u\in V$ and $u_h \in V_h$. By substituting these three inequalities in (1.19), we obtain :

$$(g,u-u_h) = a(u-u_h,\phi_g-\phi_h) + b(\phi_g-\phi_h,\lambda-\mu_h) + b(u-u_h,\xi_g-\xi_h).$$

Hence

$$|(g,u-u_h)| \leq C\left(\|u-u_h\|_X + \|\lambda-\mu_h\|_M\right)\left(\|\phi_g-\phi_h\|_X + \|\xi_g-\xi_h\|_M\right),$$

where $C = \sup(\|a\|, \|b\|)$. ∎

§ 2. A FIRST METHOD FOR SOLVING THE STOKES PROBLEM

We construct here a finite element method based on the formulation of the Stokes problem developed in §5, Chapter I. The theoretical analysis is carried as an application of the theory presented in the preceding paragraph.

2.1. THE GENERAL APPROXIMATION

We recall that Ω is a bounded domain of $\mathbb{R}^n$ (i.e. an open and connected subset) with a Lipschitz continuous boundary Γ ; we assume that $\vec{f}$ is a given function of $(H^{-1}(\Omega))^n$. Then, the homogeneous Stokes problem :

Find $(\vec{u},p)$ in $(H_0^1(\Omega))^n \times L_0^2(\Omega)$ such that

$$\text{(2.1)} \qquad \left.\begin{array}{r} -\nu\Delta\vec{u} + \overrightarrow{\text{grad}}\, p = \vec{f} \\ \text{div}\,\vec{u} = 0 \end{array}\right\} \text{in } \Omega,$$

has a unique solution.

As in §5, Chapter I, we set

$$\text{(2.2)} \begin{cases} \text{(a)} & a(\vec{u},\vec{v}) = 2\nu \sum_{i,j=1}^{n} \left(D_{ij}(\vec{u}), D_{ij}(\vec{v})\right) \\ & \text{or} \\ \text{(b)} & a(\vec{u},\vec{v}) = \nu(\overrightarrow{\text{grad}}\,\vec{u},\ \overrightarrow{\text{grad}}\,\vec{v}), \end{cases}$$

$$b(\vec{v},q) = -(q,\text{div}\,\vec{v}),$$

$$\chi = 0, \quad \langle \ell,\vec{v}\rangle = \langle \vec{f},\vec{v}\rangle .$$

For each h, let W_h and M_h be two finite-dimensional spaces such that :

$$W_h \subset H^1(\Omega), \quad M_h \subset L_0^2(\Omega).$$

Then we define

$$\text{(2.3)} \qquad W_{h,0} = W_h \cap H_0^1(\Omega), \quad X_h = (W_{h,0})^n.$$

With these spaces, problem (2.1) is approximated by :

$$(Q_h) \begin{cases} \text{Find a pair } (\vec{u}_h,p_h) \text{ in } X_h \times M_h \text{ satisfying} \\ \text{(2.4)} \quad a(\vec{u}_h,\vec{v}_h) - (p_h,\text{div}\,\vec{v}_h) = \langle \vec{f},\vec{v}_h\rangle \quad \forall \vec{v}_h \in X_h, \\ \text{(2.5)} \quad (q_h,\text{div}\,\vec{u}_h) = 0 \quad \forall q_h \in M_h. \end{cases}$$

The space V_h defined by (1.4) is :

$$V_h = \{\vec{v}_h \in X_h \ ;\ (q_h,\text{div}\,\vec{v}_h) = 0 \quad \forall q_h \in M_h\}.$$

<u>Remark 2.1</u>.

As mentioned in §1, $V_h \not\subset V$; that is, the velocities of V_h have, in general, a non-vanishing divergence. As a consequence, formulas (2.2a) and (2.2b) are *no longer equivalent*. ■

The problem (P_h) associated with (Q_h) is :

$$(P_h)\begin{cases}\text{Find } \vec{u}_h \in V_h \text{ such that} \\ (2.6) \qquad a(\vec{u}_h,\vec{v}_h) = \langle \vec{f},\vec{v}_h \rangle \quad \forall \vec{v}_h \in V_h.\end{cases}$$

In order to study the solution of problems (P_h) and (Q_h), we relate the continuous and discrete spaces by the following hypotheses.

Hypothesis H1

There exists a mapping $r_h \in \mathcal{L}\left((H^2(\Omega))^n;W_h^n\right) \cap \mathcal{L}\left((H^2(\Omega)\cap H_0^1(\Omega))^n;X_h\right)$ and an integer ℓ such that :

$$(2.7) \qquad \left(q_h, \operatorname{div}(\vec{v}-r_h\vec{v})\right) = 0 \quad \forall q_h \in M_h, \quad \forall \vec{v} \in \left(H^2(\Omega)\right)^n,$$

$$(2.8) \qquad \| r_h\vec{v}-\vec{v}\|_{1,\Omega} \leq C\,h^m \|\vec{v}\|_{m+1,\Omega} \quad \forall \vec{v} \in \left(H^{m+1}(\Omega)\right)^n,$$
$$\forall m \in \mathbb{N} \text{ with } 1 \leq m \leq \ell. \qquad \blacksquare$$

Hypothesis H2

The orthogonal projection operator ρ_h from $L_0^2(\Omega)$ onto M_h satisfies :

$$(2.9) \qquad \| q-\rho_h q\|_{0,\Omega} \leq C\,h^m \| q\|_{m,\Omega} \quad \forall q \in H^m(\Omega)\cap L_0^2(\Omega), \quad 0 \leq m \leq \ell. \qquad \blacksquare$$

THEOREM 2.1.

Under hypotheses H1 and H2, problem (P_h) has exactly one solution $\vec{u}_h$ in V_h and

$$(2.10) \qquad \lim_{h\to 0} |\vec{u}_h - \vec{u}|_{1,\Omega} = 0.$$

Moreover, if $(\vec{u},p) \in \left(H^{m+1}(\Omega)\right)^n \times \left(H^m(\Omega)\cap L_0^2(\Omega)\right)$, we have the error bound :

$$(2.11) \qquad |\vec{u}-\vec{u}_h|_{1,\Omega} \leq C\,h^m\left(\|\vec{u}\|_{m+1,\Omega} + \|p\|_{m,\Omega}\right) \text{ for } 1 \leq m \leq \ell.$$

Proof

1°/ Let us check the hypotheses of Theorem 1.1. First, $V_h \neq \{\vec{0}\}$ by virtue of (2.7). Next, we examine the ellipticity of the form $a(\cdot,\cdot)$. When a is defined by (2.2b), then

$$a(\vec{v}_h,\vec{v}_h) = \nu|\vec{v}_h|^2_{1,\Omega} \quad \forall \vec{v}_h \in X_h.$$

When a is defined by (2.2a) we make use of Korn's inequality (cf. Duvaut & Lions [22]) :

$$\sum_{i,j=1}^{n} \| D_{ij}(\vec{v}) \|^2_{0,\Omega} \geq \alpha_0 |\vec{v}|^2_{1,\Omega}, \ \alpha_0 > 0, \quad \forall \vec{v} \in (H^1_0(\Omega))^n.$$

Hence, with either definition, the form a is uniformly elliptic on X_h :

$$a(\vec{v}_h, \vec{v}_h) \geq \alpha^\star |v_h|^2_{1,\Omega} \quad \forall \vec{v}_h \in X_h. \tag{2.12}$$

Therefore, problem (P_h) has a unique solution $\vec{u}_h$ in V_h.

2°/ Now we turn to the convergence of $\vec{u}_h$. Let $\mathcal{V} = V \cap (H^2(\Omega))^n$; $\mathcal{V}$ is dense in V according to Theorem 3.8, Chapter I. Let r_h be the mapping of hypothesis H1. Clearly (2.7) implies that

$$(q_h, \operatorname{div} r_h \vec{v}) = 0 \quad \forall q_h \in M_h, \ \forall \vec{v} \in \mathcal{V};$$

that is, $r_h \vec{v} \in V_h \quad \forall \vec{v} \in \mathcal{V}$. Moreover, by (2.8) with $m = 1$, we have :

$$\| r_h \vec{v} - \vec{v} \|_{1,\Omega} \leq C_1 h \| \vec{v} \|_{2,\Omega} \quad \forall \vec{v} \in \mathcal{V}. \tag{2.13}$$

Similarly, let $\mathcal{M} = H^1(\Omega) \cap L^2_0(\Omega)$. Clearly, $\mathcal{M}$ is dense in $L^2_0(\Omega)$ because every function q of $L^2_0(\Omega)$ has the expression $q = p - \frac{1}{\text{meas}(\Omega)}(p,1)$ for some p in $L^2(\Omega)$. Also, hypothesis H2 with $m = 1$ implies that

$$\| q - \rho_h q \|_{0,\Omega} \leq C_2 h \| q \|_{1,\Omega} \quad \forall q \in \mathcal{M}. \tag{2.14}$$

Then (2.10) follows from Corollary 1.1, (2.13) and (2.14).

3°/ When $(\vec{u},p) \in (H^{m+1}(\Omega))^n \times (H^m(\Omega) \cap L^2_0(\Omega))$ for $1 \leq m \leq \ell$, then (2.11) follows immediately from (2.8), (2.9) and (1.7). ∎

In order to derive a sharper estimate for $\| \vec{u} - \vec{u}_h \|_{0,\Omega}$, we put problem (P_h) in the setting of Theorem 1.2. We take $H = (L^2(\Omega))^n$; for $\vec{g}$ in $(L^2(\Omega))^n$, problem (1.17) is the homogeneous Stokes problem :

$$(2.15) \quad \begin{cases} \text{Find } (\vec{\phi}_{\vec{g}}, \xi_{\vec{g}}) \text{ in } (H^1_0(\Omega))^n \times L^2_0(\Omega) \text{ such that :} \\ \nu(\overrightarrow{\text{grad}}\, \vec{v}, \overrightarrow{\text{grad}}\, \vec{\phi}_{\vec{g}}) - (\xi_{\vec{g}}, \operatorname{div} \vec{v}) = (\vec{g}, \vec{v}) \quad \forall \vec{v} \in (H^1_0(\Omega))^n, \\ \qquad (q, \operatorname{div} \vec{\phi}_{\vec{g}}) = 0 \quad \forall q \in L^2_0(\Omega). \end{cases}$$

In other words, $(\vec{\phi}_{\vec{g}}, \xi_{\vec{g}})$ satisfies

$$(2.15') \quad \begin{cases} \left. \begin{aligned} -\nu \Delta \vec{\phi}_{\vec{g}} + \overrightarrow{\text{grad}}\, \xi_{\vec{g}} &= \vec{g} \\ \operatorname{div} \vec{\phi}_{\vec{g}} &= 0 \end{aligned} \right\} \text{ in } \Omega, \\ \qquad \vec{\phi}_{\vec{g}}|_\Gamma = \vec{0}. \end{cases}$$

DEFINITION 2.1.

We say that problem (2.15') is *regular* if the mapping $(\vec{\phi}_{\vec{g}}, \xi_{\vec{g}}) \mapsto -\nu\Delta\vec{\phi}_{\vec{g}} + \overrightarrow{\text{grad}}\, \xi_{\vec{g}}$ is an isomorphism from $[(H^2(\Omega))^n \cap V] \times [H^1(\Omega) \cap L^2_0(\Omega)]$ *onto* $(L^2(\Omega))^n$, i.e. if

$$\|\vec{\phi}_{\vec{g}}\|_{2,\Omega} + \|\xi_{\vec{g}}\|_{1,\Omega} \leq C\|\vec{g}\|_{0,\Omega}. \tag{2.16}$$

According to Theorem 5.2, Chapter I, we know that problem (2.15') is regular if Γ is of class $\mathcal{C}^2$. When Γ is only Lipschitz continuous - and subsequently Γ will be a polygonal line - Theorem 5.2 asserts that this problem is regular provided Ω is a plane, bounded and convex domain.

THEOREM 2.2.

We assume that hypotheses H1 and H2 are satisfied and that problem (2.15') is regular. Then, if $(\vec{u},p) \in (H^{m+1}(\Omega))^n \times (H^m(\Omega) \cap L^2_0(\Omega))$, we have the following error bound :

$$\|\vec{u}-\vec{u}_h\|_{0,\Omega} \leq C\, h^{m+1} \left(\|\vec{u}\|_{m+1,\Omega} + \|p\|_{m,\Omega}\right) \quad \text{for } 1 \leq m \leq \ell. \tag{2.17}$$

Proof.

Let us apply Theorem 1.2 :

$$\|\vec{u}-\vec{u}_h\|_{0,\Omega} \leq C\{|\vec{u}-\vec{u}_h|_{1,\Omega} + \inf_{q_h\in M_h} \|p-q_h\|_{0,\Omega}\} \times$$

$$\times \sup_{\vec{g}\in(L^2(\Omega))^n} \left\{\frac{1}{\|\vec{g}\|_{0,\Omega}} \left(\inf_{\vec{\phi}_h\in V_h} |\vec{\phi}_{\vec{g}}-\vec{\phi}_h|_{1,\Omega} + \inf_{\xi_h\in M_h} \|\xi_{\vec{g}}-\xi_h\|_{0,\Omega}\right)\right\}.$$

As problem (2.15') is regular, $\vec{\phi}_{\vec{g}} \in (H^2(\Omega))^n \cap V$ and $\xi_{\vec{g}} \in H^1(\Omega) \cap L^2_0(\Omega)$. Therefore, by virtue of H1 and H2, $r_h\vec{\phi}_{\vec{g}} \in V_h$ and

$$\|\vec{\phi}_{\vec{g}} - r_h\vec{\phi}_{\vec{g}}\|_{1,\Omega} \leq C\, h\|\vec{\phi}_{\vec{g}}\|_{2,\Omega}\, , \|\xi_{\vec{g}}-\rho_h\xi_{\vec{g}}\|_{0,\Omega} \leq C\, h\|\xi_{\vec{g}}\|_{1,\Omega}.$$

Therefore, by (2.16)

$$\|\vec{u}-\vec{u}_h\|_{0,\Omega} \leq C\, h\left(|\vec{u}-\vec{u}_h|_{1,\Omega} + \inf_{q_h\in M_h} \|p-q_h\|_{0,\Omega}\right).$$

Then (2.11) and (2.9) yield (2.17). ■

We end this section by establishing the convergence of p_h. In order to check the inf-sup condition, we require an additional hypothesis.

Hypothesis H3

For each q_h in M_h, there exists a function $\vec{v}_h$ in X_h $\left(=(W_{h,0})^n\right)$, such that

$$(2.18) \qquad (\operatorname{div} \vec{v}_h - q_h, s_h) = 0, \quad \forall s_h \in M_h,$$

$$(2.19) \qquad |\vec{v}_h|_{1,\Omega} \leq C \| q_h \|_{0,\Omega}. \qquad \blacksquare$$

This hypothesis is a discrete analogue of Lemma 3.2, Chapter I : the divergence operator is an isomorphism from $V^{\perp}$ onto $L^2_0(\Omega)$.

THEOREM 2.3.

Under hypotheses H1, H2 and H3, problem (Q_h) has exactly one solution $(\vec{u}_h, p_h) \in V_h \times M_h$, where $\vec{u}_h$ is *the* solution of (P_h) and

$$(2.20) \qquad \lim_{h\to o}(|\vec{u}-\vec{u}_h|_{1,\Omega} + \| p-p_h \|_{0,\Omega}) = 0.$$

Moreover, if $(\vec{u},p) \in \left(H^{m+1}(\Omega)\right)^n \times \left(H^m(\Omega) \cap L^2_0(\Omega)\right)$, the following error bound holds :

$$(2.21) \qquad |\vec{u}-\vec{u}_h|_{1,\Omega} + \| p-p_h \|_{0,\Omega} \leq C\, h^m (\| \vec{u} \|_{m+1,\Omega} + \| p \|_{m,\Omega}) \quad \text{for } 1 \leq m \leq \ell .$$

Proof

Let us check the inf-sup condition for $\vec{v}_h$ in X_h and q_h in M_h. By applying H3 to any q_h of M_h, we find that there exists $\vec{v}_h$ in X_h with :

$$(q_h, \operatorname{div} \vec{v}_h) = \| q_h \|^2_{0,\Omega} \geq \frac{1}{C} \| q_h \|_{0,\Omega} |\vec{v}_h|_{1,\Omega} .$$

Hence,

$$\sup_{\vec{v}_h \in X_h} \frac{(q_h, \operatorname{div} \vec{v}_h)}{|\vec{v}_h|_{1,\Omega}} \geq \frac{1}{C} \| q_h \|_{0,\Omega} \quad \forall q_h \in M_h.$$

Together with Theorems 2.1 and 1.1, this implies that there exists a unique p_h in M_h such that $(\vec{u}_h, p_h)$ is the only solution of (Q_h), where $\vec{u}_h$ is the solution of (P_h). Furthermore, by virtue of (1.13) :

$$\| p-p_h \|_{0,\Omega} \leq C\{ |\vec{u}-\vec{u}_h|_{1,\Omega} + \inf_{q_h \in M_h} \| p-q_h \|_{0,\Omega} \}.$$

Then owing to (2.10) and H2 with $m = 1$, we get :

$$\lim_{h\to o} \| p-p_h \|_{0,\Omega} = 0.$$

Similarly, (2.21) follows from H2 with $m \leq \ell$ and (2.11). $\blacksquare$

2.2. EXAMPLE 1 : A FIRST-ORDER APPROXIMATION

Let us first recall the notations and the most important facts, for our purpose, about the classical finite element approximation. For detailed proofs and further material, the reader can refer to Ciarlet [14].

DEFINITION 2.2.

For each integer $m \geq 0$, we denote by P_m the space of all polynomials, defined on $\mathbb{R}^n$, of degree less than or equal to m. We denote by Q_m the space of polynomials spanned by $\prod_{i=1}^{n} x_i^{\alpha_i}$ with $0 \leq \alpha_i \leq m$. ∎

To simplify the discussion, we shall assume in the examples that Ω is a *plane polygonal domain*.

For each $h > 0$, let $\mathcal{T}_h$ be a triangulation of $\bar{\Omega}$ made of closed triangles K with diameters bounded by h. In other words :

$$\bar{\Omega} = \bigcup_{K \in \mathcal{T}_h} K,$$

where any two triangles K_1 and K_2 are either disjoint or share at most one side or one vertex. The size and shape of each triangle K are specified by two quantities :

$$h_K = \text{diameter of } K, \; h_K \leq h,$$

and

$$\rho_K = \text{ diameter of the inscribed circle in } K.$$

The regularity of a triangle K is measured by the ratio

$$\sigma_K = h_K / \rho_K.$$

DEFINITIONS 2.3

1°/ A family $\mathcal{T}_h$ of triangulations of $\bar{\Omega}$ is said to be regular as h tends to zero if there exists a number $\sigma > 0$, independent of h and K, such that

$$\sigma_K \leq \sigma \quad \forall K \in \mathcal{T}_h.$$

2°/ In addition, $\mathcal{T}_h$ is said to be uniformly regular as h tends to zero if there exists another constant $\tau > 0$ such that

$$\tau h \leq h_K \leq \sigma \rho_K \quad \forall K \in \mathcal{C}_h. \qquad \blacksquare$$

We denote by $\hat{K}$ the unit reference triangle in the reference $(\hat{x}_1,\hat{x}_2)$-space with vertices $\hat{a}_1 = (0,0)$, $\hat{a}_2 = (1,0)$ and $\hat{a}_3 = (0,1)$. If K is any triangle with vertices a_1, a_2 and a_3 (cf. figure 3) there exists exactly one affine mapping

$$F_K(\hat{x}) = B_K\hat{x} + b_K \tag{2.22}$$

that maps $\hat{K}$ onto K with $F_K(\hat{a}_i) = a_i$ for $1 \leq i \leq 3$. Furthermore, it can be easily shown that the matrix B_K is non singular and satisfies the following bounds :

$$\| B_K \| \leq C_1 h_K, \quad \| B_K^{-1} \| \leq C_2/\rho_K, \tag{2.23}$$

where $\| \cdot \|$ stands for the matrix norm associated with the Euclidean norm of $\mathbb{R}^2$. The composition with F_K is indicated by a "hat" : $\hat{v} = v \circ F_K$. According to convenience, we shall sometimes replace the Euclidean coordinates by the barycentric coordinates $\lambda_1,\lambda_2,\lambda_3$ defined by :

$$\lambda_i \in P_1, \quad \sum_{i=1}^{3} \lambda_i \equiv 1, \quad \lambda_i(a_j) = \delta_i^j \text{ for } 1 \leq i,j \leq 3. \tag{2.24}$$

LEMMA 2.1.

For each integer $m \geq 0$ and for all real p with $1 \leq p \leq \infty$, the mapping $v \mapsto \hat{v} = v \circ F_K$ is an isomorphism from $W^{m,p}(K)$ onto $W^{m,p}(\hat{K})$ and the following bounds hold :

$$|\hat{v}|_{m,p,\hat{K}} \leq C_1 \| B_K \|^m |\det B_K|^{-1/p} |v|_{m,p,K} \quad \forall v \in W^{m,p}(K), \tag{2.25}$$

$$|v|_{m,p,K} \leq C_2 \| B_K^{-1} \|^m |\det B_K|^{1/p} |\hat{v}|_{m,p,\hat{K}} \quad \forall \hat{v} \in W^{m,p}(\hat{K}). \tag{2.26}$$

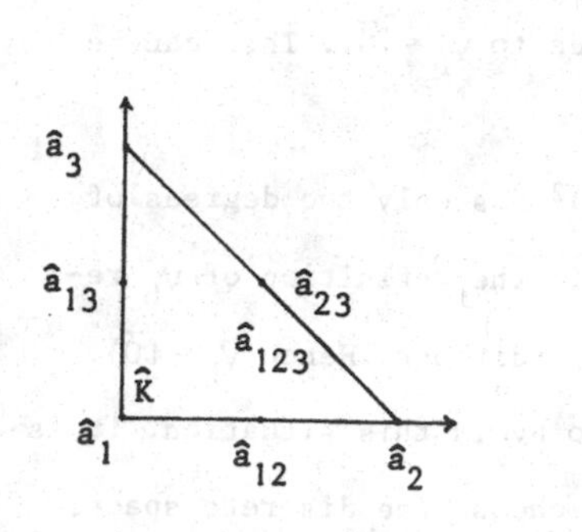

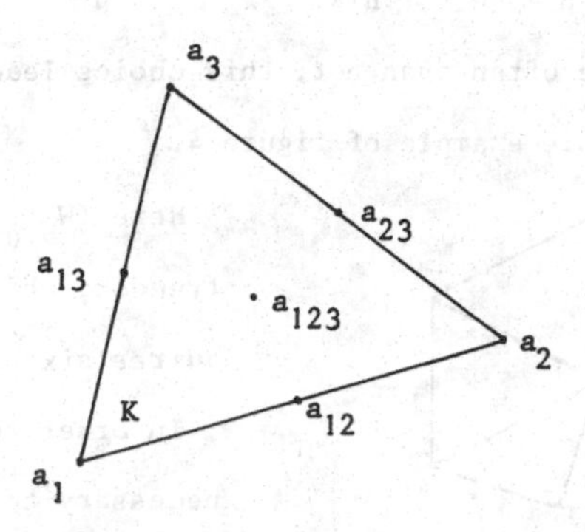

FIGURE 3

THEOREM 2.4.

Let k and m be two integers with $0 \leq m \leq k+1$. Let $\hat{\Pi}$ be an operator in $\mathcal{L}(W^{k+1,p}(\hat{K})\,;\,W^{m,p}(\hat{K}))$ and let Π be the operator of $\mathcal{L}(W^{k+1,p}(K)\,;\,W^{m,p}(K))$ defined by

$$(\Pi v)\circ F_K = \hat{\Pi}(v \circ F_K) \quad (\text{i.e. } \widehat{\Pi v} = \hat{\Pi}\hat{v}). \tag{2.27}$$

If P_k is *invariant* under $\hat{\Pi}$, i.e.

$$\hat{\Pi}p = p \quad \forall p \in P_k,$$

then there exists a constant $\hat{C}$, independent of v and K, such that :

$$|v-\Pi v|_{m,p,K} \leq \hat{C}\,\|B_K\|^{k+1}\,\|B_K^{-1}\|^m\,|v|_{k+1,p,K} \quad \forall v \in W^{k+1,p}(K). \tag{2.28}$$

COROLLARY 2.1.

Under the hypotheses of Theorem 2.4 there exists a constant C, independent of Π, v and K such that

$$|v-\Pi v|_{m,p,K} \leq C\,h_K^{k+1}\,\rho_K^{-m}\,|v|_{k+1,p,K} \quad \forall v \in W^{k+1,p}(K). \tag{2.29}$$

COROLLARY 2.2.

Let $\partial\hat{K}$ denote the boundary of $\hat{K}$. If P_m is invariant under $\hat{\Pi} \in \mathcal{L}(H^{m+1}(\hat{K})\,;\,L^2(\partial\hat{K}))$, then there exists a constant $\hat{C}$ such that

$$\|\hat{\Pi}\hat{v}-\hat{v}\|_{0,\partial\hat{K}} \leq \hat{C}\,|\hat{v}|_{m+1,\hat{K}} \quad \forall \hat{v} \in H^{m+1}(\hat{K}).$$

The most straightforward choice of discrete spaces is :

$$W_h = \{w_h \in \mathcal{C}^0(\bar{\Omega})\,;\, w_h|_K \in P_1 \quad \forall K \in \mathcal{T}_h\} \subset W^{1,\infty}(\Omega),$$

$$\mathcal{O}_h = \{q_h \in L^2(\Omega)\,;\, q_h|_K \in P_0 \quad \forall K \in \mathcal{T}_h\}, \quad M_h = \mathcal{O}_h \cap L_0^2(\Omega). \tag{2.30a}$$

Unfortunately, more often than not, this choice leads to $V_h = \{\vec{0}\}$. This can be checked in the simple example of figure 4.

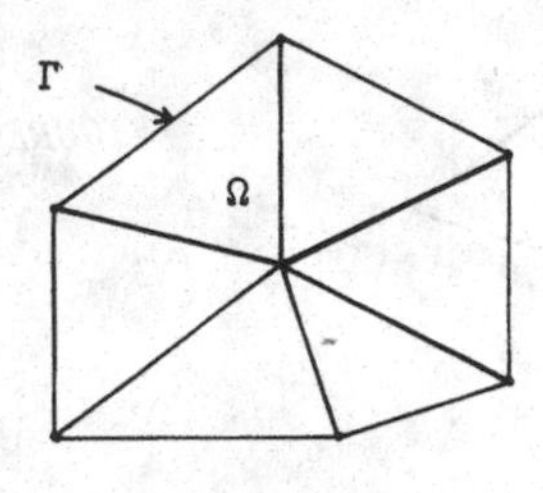

FIGURE 4

Here $(W_{h,0})^2$ has only two degrees of freedom, while the definition of V_h requires six conditions. Hence $V_h = \{\vec{0}\}$.

In order to avoid this situation, it is necessary to choose the discrete spaces with more care. Since it appears from the above example, that $W_{h,0}$ requires more degrees of freedom, it seems reasonable to choose :

(2.30b) $$W_h = \{w_h \in \mathcal{C}^\circ(\bar{\Omega}) \ ; \ w_h|_K \in P_2 \ \ \forall K \in \mathcal{T}_h\} \subset W^{1,\infty}(\Omega),$$

and M_h unchanged. As far as the degrees of freedom are concerned, we can choose the values of the functions $w_h \in W_h$ (resp. $q_h \in \mathcal{O}_h$) at the vertices a_i and at the midpoints a_{ij} of the sides of K (resp. at the centroid a_{123} of K), as in figure 3.

For this choice, let us check hypotheses H1, H2 and H3 with $\ell = 1$. We start by constructing the operator r_h.

LEMMA 2.2.

Let K be an element of $\mathcal{T}_h$ as in figure 3. For each function v of $\mathcal{C}^\circ(\bar{K})$, there exists one and only one function $\Pi_K v \in P_2$ defined by :

(2.31) $$\begin{cases} \Pi_K v(a_i) = v(a_i) & 1 \leq i \leq 3, \\ \displaystyle\int_{[a_i,a_j]} (\Pi_K v - v)\, d\sigma = 0 & 1 \leq i < j \leq 3, \end{cases}$$

where $[a_i,a_j]$ denotes the side of K with end points a_i and a_j. Moreover, on $[a_i,a_j]$, $\Pi_K v$ depends only upon the values of v on this side.

Proof

Every polynomial p of P_2 on K is of the form

(2.32) $$p \equiv \sum_{i=1}^{3} p(a_i)\lambda_i + 4\Big(\sum_{1\leq i<j\leq 3} p(a_{ij})\lambda_i\lambda_j\Big)$$

Thus (2.31) is a system of six linear equations with six unknowns. Therefore we must show that $v = 0 \Rightarrow \Pi_K v = 0$. Owing to (2.31) and (2.32), it suffices to show that $\Pi_K v(a_{ij}) = 0$ for $1 \leq i < j \leq 3$.

On $[a_i,a_j]$, $\Pi_K v \equiv 4\Pi_K v(a_{ij})\lambda_i\lambda_j$, and according to (2.31)

$$\int_{[a_i,a_j]} \Pi_K v \, d\sigma = 0.$$

As $\displaystyle\int_{[a_i,a_j]} \lambda_i\lambda_j \, d\sigma > 0$, this implies that $\Pi_K v(a_{ij}) = 0$. ■

Clearly P_2 is invariant under Π_K.

LEMMA 2.3.

Each function $v \in H^i(K)$ for $i = 2$ or 3 satisfies :

(2.33) $$|v - \Pi_K v|_{j,K} \leq C\, h_K^i \rho_K^{-j} |v|_{i,K} \quad \text{for } 0 \leq j \leq i,$$

where the constant C is independent of v and K.

Proof

When the change of variable (2.22) is applied to (2.31), we get :

$$(\Pi_K v \circ F_K)(\hat{a}_k) = (v \circ F_K)(\hat{a}_k) \quad 1 \leq k \leq 3 \ ,$$

$$\int_{[\hat{a}_k, \hat{a}_\ell]} (\Pi_K v - v) \circ F_K \, d\hat{\sigma} = 0 \quad 0 \leq k < \ell \leq 3 \ .$$

In other words :

$$\widehat{\Pi_K v} = \Pi_{\hat{K}} \hat{v}.$$

Besides that, P_{i-1} (at least) is invariant under $\Pi_{\hat{K}}$. Therefore (2.33) follows from (2.29) with $p = 2$, $m = j$ and $k = i-1$. ∎

Now, we are in a position to define the operator r_h :

For each $\vec{v} \in (\mathcal{C}^\circ(\bar{\Omega}))^2$, we set

(2.34) $$(r_h \vec{v})_i\big|_K = \Pi_K v_i \quad i = 1,2 \quad \forall K \in \mathcal{T}_h.$$

This defines a function of $(\mathcal{C}^\circ(\bar{\Omega}))^2$ since $\Pi_K v\big|_{[a_i, a_j]}$ depends solely upon $v\big|_{[a_i, a_j]}$. Since Ω is a polygon, we have $r_h \vec{v}\big|_\Gamma = \vec{0}$ whenever $\vec{v}$ vanishes on Γ. Therefore $r_h \in \mathcal{L}\big((H^2(\Omega))^2 \ ; W_h^2\big) \cap \mathcal{L}\big((H^2(\Omega) \cap H_0^1(\Omega)\ )^2 ; X_h\big)$.

LEMMA 2.4.

1°/ r_h satisfies (2.7).

2°/ If $\mathcal{T}_h$ is a regular family of triangulations of $\bar{\Omega}$, then r_h satisfies (2.8) for $\ell = 2$.

Proof

1°/ For each $q_h \in M_h$, we have :

$$(\operatorname{div} r_h \vec{v}, q_h) = \sum_{K \in \mathcal{T}_h} q_h\big|_K \int_K \operatorname{div} r_h \vec{v} \, dx.$$

From (2.34) and (2.31) we derive :

$$\int_K \operatorname{div} r_h\vec{v}\,dx = -\int_{\partial K} \Pi_K\vec{v}\cdot\vec{\nu}\,d\sigma = -\int_{\partial K} \vec{v}\cdot\vec{\nu}\,d\sigma = \int_K \operatorname{div}\vec{v}\,dx.$$

Therefore

$$(\operatorname{div} r_h\vec{v}, q_h) = (\operatorname{div}\vec{v}, q_h) \quad \forall q_h \in M_h.$$

2°/ Let us apply Lemma 2.3 to $\vec{v} \in \left(H^i(\Omega)\right)^2$ for $i = 2$ or 3 :

$$\|\vec{v}-r_h\vec{v}\|^2_{1,\Omega} = \sum_{K\in\mathcal{T}_h} \|\vec{v}-\Pi_k\vec{v}\|^2_{1,K} \leq C^2\sigma_K^2\, h_K^{2(i-1)} \|\vec{v}\|^2_{i,\Omega}.$$

Then the regularity of $\mathcal{T}_h$ implies that

$$\|\vec{v}-r_h\vec{v}\|_{1,\Omega} \leq C\sigma\, h^{i-1}\|\vec{v}\|_{i,\Omega}. \qquad \blacksquare$$

Thus, if $\mathcal{T}_h$ is regular, our operator r_h satisfies H1.

Next, let us examine the orthogonal projection operator ρ_h on $\mathcal{O}_h$. Because the functions of $\mathcal{O}_h$ have *no continuity requirement* across the interelement boundaries, we see that

$\rho_h q|_K$ = orthogonal projection ϖ_K, in $L^2(K)$, of q on P_0, i.e.

$$\rho_h q|_K = \varpi_K q = \frac{1}{\operatorname{meas}(K)}\int_K q(x)\,dx.$$

From this definition, we derive immediately that ρ_h projects $L^2_0(\Omega)$ onto M_h. Clearly $\bar{\omega}_K$ is invariant under an affine transformation and P_0 is invariant under ϖ_K. Therefore Corollary 2.1 implies that

$$\|q-\rho_h q\|_{0,\Omega} \leq C\,h\|q\|_{1,\Omega} \quad \forall q \in H^1(\Omega).$$

Hence hypothesis H2 is valid.

Finally, we turn to hypothesis H3 that deals with the inf-sup condition. Let $q_h \in M_h$; according to Lemma 3.2, Chapter 1, there exists exactly one function $\vec{v} \in V^{\perp}$ such that

$$\operatorname{div}\vec{v} = q_h, \quad |\vec{v}|_{1,\Omega} \leq C\|q_h\|_{0,\Omega}.$$

In order to construct a function $\vec{v}_h$ satisfying H3, the simplest thing would be to take $\vec{v}_h = r_h\vec{v}$. However, this is not possible because $\vec{v}$ is not necessarily continuous. This difficulty can be overcome by first taking the orthogonal projection $\vec{w}_h$ of $\vec{v}$ on $(W_{h,0})^2$ for the scalar product of $\left(H^1_0(\Omega)\right)^2$:

(2.35) $$a(\vec{w}_h-\vec{v},\vec{z}_h)=0 \quad \forall \vec{z}_h \in (W_{h,0})^2.$$

Then on each triangle K we define the function $\vec{v}_h$ of $(W_{h,0})^2$ by :

(2.36) $$\begin{cases} \vec{v}_h(a_i)=\vec{w}_h(a_i) & \text{for } 1 \leq i \leq 3 \\ \displaystyle\int_{[a_i,a_j]} (\vec{v}_h-\vec{v})\,d\sigma = \vec{0} & \text{for } 1 \leq i < j \leq 3. \end{cases}$$

Clearly the second equation of (2.36) makes sure that

$$\int_K (\operatorname{div} \vec{v}_h - q_h)\,dx = 0 \quad \forall K \in \mathcal{T}_h.$$

Hence $\vec{v}_h$ satisfies (2.18). The estimate (2.19) is established by the next lemma.

<u>LEMMA 2.5.</u>

<u>Suppose that Ω is convex. Let $\mathcal{T}_h$ be a uniformly regular family of triangulations of $\bar{\Omega}$. Then the function $\vec{v}_h$ defined by (2.35) and (2.36) satisfies the bound</u> :

(2.37) $$|\vec{v}_h|_{1,\Omega} \leq C \| q_h \|_{0,\Omega}.$$

<u>Proof</u>

We set $\vec{e}_h = \vec{v}_h - \vec{w}_h \in (W_{h,0})^2$ and $\vec{e} = \vec{v} - \vec{w}_h$. Then

$$|\vec{v}_h|_{1,\Omega} \leq |\vec{w}_h|_{1,\Omega} + |\vec{e}_h|_{1,\Omega} \leq |\vec{v}|_{1,\Omega} + |\vec{e}_h|_{1,\Omega}.$$

Thus it suffices to estimate $|\vec{e}_h|_{1,\Omega}$. First, note that

(2.38) $$\vec{e}_h(a_i) = \vec{0} \quad \text{for } 1 \leq i \leq 3, \qquad \int_{[a_i,a_j]} (\vec{e}_h - \vec{e})\,d\sigma = \vec{0} \quad \text{for } 1 \leq i < j \leq 3.$$

Therefore, in each triangle K, $\vec{e}_h$ is of the form

(2.39) $$\vec{e}_h = \sum_{1\leq i<j\leq 3} \vec{e}_h(a_{ij}) p_{ij},$$

where $p_{ij} = 4\lambda_i\lambda_j$.

According to (2.26) :

$$|p_{ij}|_{1,K} \leq C_1 \| B_K^{-1} \| \, |\det B_K|^{\frac{1}{2}} \, |\hat{p}_{ij}|_{1,\hat{K}} ,$$

hence

$$|p_{ij}|_{1,K} \leq C_2 \| B_K^{-1} \| \, |\det B_K|^{\frac{1}{2}} .$$

Besides that by integrating (2.39) over $[a_i,a_j]$ and using (2.38), we get :

$$\vec{e}_h(a_{ij}) = \left(\int_{[\hat{a}_i,\hat{a}_j]} \hat{p}_{ij}\, d\hat{\sigma}\right)^{-1} \int_{[\hat{a}_i,\hat{a}_j]} \hat{\vec{e}}\, d\hat{\sigma} \ .$$

Therefore

$$\|\vec{e}_h(a_{ij})\| \leq C_3 \|\hat{\vec{e}}\|_{0,\partial\hat{K}} \leq C_4 \left(\|\hat{\vec{e}}\|^2_{0,\hat{K}} + |\hat{\vec{e}}|^2_{1,\hat{K}}\right)^{\frac{1}{2}},$$

where $\|\cdot\|$ denotes the Euclidean norm of $\mathbb{R}^2$; then by (2.25), we have :

$$\|\vec{e}_h(a_{ij})\| \leq C_5 |\det(B_K)|^{-\frac{1}{2}} \left(\|\vec{e}\|^2_{0,K} + \|B_K\|^2 |\vec{e}|^2_{1,K}\right)^{\frac{1}{2}}.$$

Hence

$$|\vec{e}_h|_{1,K} \leq C_6 \|B_K^{-1}\| \left(\|\vec{e}\|^2_{0,K} + \|B_K\|^2 |\vec{e}|^2_{1,K}\right)^{\frac{1}{2}} \ .$$

As $\mathcal{T}_h$ is uniformly regular, it follows that

$$|\vec{e}_h|_{1,K} \leq C_7\sigma \left(h_K^{-2} \|\vec{e}\|^2_{0,K} + |\vec{e}|^2_{1,K}\right)^{\frac{1}{2}};$$

thus

$$|\vec{e}_h|_{1,\Omega} \leq C_8 \left(h^{-2} \|\vec{e}\|^2_{0,\Omega} + |\vec{e}|^2_{1,\Omega}\right)^{\frac{1}{2}} \ . \tag{2.40}$$

It remains to evaluate the norms of $\vec{e}$ in the right-hand side of (2.40). First, (2.35) implies that

$$|\vec{e}|_{1,\Omega} \leq |\vec{v}|_{1,\Omega}.$$

Next, the convexity of Ω implies that $-\Delta$ is an isomorphism from $H^2(\Omega) \cap H^1_0(\Omega)$ onto $L^2(\Omega)$ (cf. remark 1.1 n°2, Chapter I). Then, by using the classical duality argument :

$$\|e_i\|_{0,\Omega} = \sup_{g \in L^2(\Omega)} \frac{(e_i, g)}{\|g\|_{0,\Omega}} \ ,$$

we find

$$\|\vec{e}\|_{0,\Omega} \leq C_9 h |\vec{e}|_{1,\Omega} \ ,$$

thus proving the lemma. ∎

Since hypotheses H1, H2 and H3 are valid, we have the following result.

<u>THEOREM 2.5.</u>

<u>Let the solution $(\vec{u},p)$ of the Stokes problem satisfy : $\vec{u} \in [H^2(\Omega)]^2$ and $p \in H^1(\Omega) \cap L^2_0(\Omega)$, and let the spaces M_h and W_h be defined by (2.30).</u>

1°/ <u>If the family $\mathcal{T}_h$ is regular, we have the following estimate :</u>

(2.41) $$|\vec{u}-\vec{u}_h|_{1,\Omega} \leq C_1 h(\|\vec{u}\|_{2,\Omega} + \|p\|_{1,\Omega}).$$

2°/ Moreover, if Ω is convex, then

(2.42) $$\|\vec{u}-\vec{u}_h\|_{0,\Omega} \leq C_2 h^2(\|\vec{u}\|_{2,\Omega} + \|p\|_{1,\Omega}).$$

3°/ If, besides that, the family $\mathcal{T}_h$ is uniformly regular, then

(2.43) $$\|p-p_h\|_{0,\Omega} \leq C_3 h(\|\vec{u}\|_{2,\Omega} + \|p\|_{1,\Omega}).$$

2.3. EXAMPLE 2 : A SECOND-ORDER APPROXIMATION

The method described in the previous section is disappointing in that it requires finite elements of degree two for the velocity in order to obtain barely a first-order approximation. A closer look at the error estimates shows that the loss of precision arises from the coarseness of the space M_h which yields a poor approximation of the incompressibility condition. In this section, we shall derive a second-order method by slightly modifying the spaces W_h and M_h.

Again let K be an element of $\mathcal{T}_h$ as in figure 3. let P_k be the space of polynomials spanned by :

$$\lambda_1^2,\ \lambda_2^2,\ \lambda_3^2,\ \lambda_1\lambda_2,\ \lambda_2\lambda_3,\ \lambda_1\lambda_3,\ \lambda_1\lambda_2\lambda_3.$$

Clearly $P_2 \subset P_K \subset P_3$. Moreover, P_K is Σ_K-unisolvent*, where

$$\Sigma_K = \{a_i\}_{1\leq i\leq 3} \cup \{a_{ij}\}_{1\leq i<j\leq 3} \cup \{a_{123}\}$$

and its basis functions are :

$$p_i = \lambda_i(2\lambda_i - 1) + 3\lambda_1\lambda_2\lambda_3,\quad 1 \leq i \leq 3,$$

$$p_{ij} = 4\lambda_i\lambda_j - 12\lambda_1\lambda_2\lambda_3\quad,\ 1 \leq i < j \leq 3,$$

$$p_{123} = 27\lambda_1\lambda_2\lambda_3.$$

We take the following spaces

---

*Let $S = \{a_i\}$ be a set of N distinct points of $\mathbb{R}^n$. An N-dimensional space P of real-valued functions defined on S is said to be S-unisolvent if for every set of N real numbers $\{\alpha_i\}$ there exists exactly one function $p \in P$ such that

$$p(a_i) = \alpha_i,\quad 1 \leq i \leq N.$$

$$(2.44)\quad \begin{cases} W_h = \{w_h \in \mathcal{C}^\circ(\bar{\Omega})\ ;\ w_h|_K \in P_K \quad \forall K \in \mathcal{T}_h\} \\ \mathcal{O}_h = \{q_h \in L^2(\Omega)\ ;\ q_h|_K \in P_1 \quad \forall K \in \mathcal{T}_h\},\ M_h = \mathcal{O}_h \cap L^2_0(\Omega), \end{cases}$$

with the degrees of freedom of the functions of W_h located in $\sum_K$. Again, there is no continuity requirement on the functions of $\mathcal{O}_h$.

Let us construct a restriction operator r_h.

LEMMA 2.6.

For each $K \in \mathcal{T}_h$ and for every function v of $[H^2(K)]^2$, there exists one and only one function $\Pi_K \vec{v} \in P_K^2$ defined by :

$$(2.45)\quad \begin{cases} \Pi_K \vec{v}(a_i) = \vec{v}(a_i) \text{ for } 1 \leq i \leq 3, \\ \displaystyle\int_{[a_i,a_j]} (\Pi_K \vec{v} - \vec{v})\, d\sigma = \vec{0} \text{ for } 1 \leq i < j \leq 3, \\ \displaystyle\int_K x_\ell \operatorname{div}(\Pi_K \vec{v} - \vec{v})\, dx = 0 \text{ for } \ell = 1,2. \end{cases}$$

Moreover, $\Pi_K \vec{v}|_{[a_i,a_j]}$ depends solely upon the value of $\vec{v}$ on $[a_i,a_j]$.

Proof

Let K be an element of $\mathcal{T}_h$ and $\vec{v}$ a function of $[H^2(K)]^2$. We must check that (2.45) determines $\Pi_K \vec{v}$ uniquely. First, we remark that P_K reduces to P_2 on the sides of K since $\lambda_1\lambda_2\lambda_3$ vanishes there. Therefore, as in Lemma 2.2, the first two conditions of (2.45) define uniquely $\Pi_K \vec{v}$ on the vertices and on the sides of K. Furthermore $\Pi_K \vec{v}|_{[a_i,a_j]}$ depends only upon $\vec{v}|_{[a_i,a_j]}$.

It remains to examine the value of $\Pi_K \vec{v}$ at the centroid of K. By Green's formula, the third condition of (2.45) yields :

$$(2.46)\quad \int_K (\Pi_K \vec{v})_\ell\, dx = \int_K v_\ell\, dx + \int_{\partial K} x_\ell\, \gamma_\nu (\Pi_K \vec{v} - \vec{v})\, d\sigma \quad \ell = 1,2.$$

Since the right-hand side of (2.46) consists of known quantities and since $\int_K p_{123}\, dx > 0$, it follows that (2.46) determines uniquely $(\Pi_K \vec{v})_\ell (a_{123})$, thus proving the lemma. ■

Now, we define our operator r_h by :

$$r_h\vec{v}|_K = \Pi_K\vec{v} \quad \forall K\in\mathcal{T}_h, \quad \forall\vec{v}\in \left(H^2(\Omega)\right)^2.$$

Clearly, $r_h\in\mathcal{L}\left(\left(H^2(\Omega)\right)^2 \; ; \; W_h^2\right)\cap\mathcal{L}\left(\left(H^2(\Omega)\cap H_0^1(\Omega)\right)^2 \; ; \; W_{h,0}^2\right)$. Furthermore, the last two conditions of (2.45) imply that, for $\vec{v}\in [H^2(\Omega)]^2$,

$$(2.47) \qquad \left(q_h, \mathrm{div}(r_h\vec{v}-\vec{v})\right) = 0 \quad \forall q_h\in\mathcal{O}_h.$$

This takes care of the first part of hypothesis H1. The next lemma deals with the second part.

LEMMA 2.7.

There exists a constant C independent of K such that

$$(2.48) \qquad |\Pi_K\vec{v}-\vec{v}|_{1,K} \leq C\sigma_K^2 h_K^k \|\vec{v}\|_{k+1,K}, \; k=1,2 \quad \forall\vec{v}\in [H^{k+1}(K)]^2 .$$

Proof

The difficulty in this proof is that, because of the third condition of (2.45), the operator Π_K is not invariant under an affine transformation. Therefore, we replace Π_K by another operator $\tilde{\Pi}_K\in\mathcal{L}\left(\left(H^2(K)\right)^2;(P_K)^2\right)$ defined by :

$$(2.45') \qquad \begin{cases} \text{the first two conditions of (2.45) unchanged,} \\ \displaystyle\int_K (\tilde{\Pi}_K\vec{v}-\vec{v})\,dx = 0. \end{cases}$$

Clearly, (2.45') defines $\tilde{\Pi}_K\vec{v}$ uniquely, $\tilde{\Pi}_K$ is invariant under an affine transformation and $(P_K)^2$ is invariant under $\tilde{\Pi}_K$. Hence, we can apply Corollary 2.1 with $m=1$ and $k=1$ or 2 :

$$(2.49) \qquad |\vec{v}-\tilde{\Pi}_K\vec{v}|_{1,K} \leq C_1\sigma_K h_K^k |\vec{v}|_{k+1,K} \quad \forall\vec{v}\in [H^{k+1}(K)]^2.$$

Now, it suffices to evaluate the difference $\Pi_K\vec{v}-\tilde{\Pi}_K\vec{v}$. Obviously, $\Pi_K\vec{v}$ and $\tilde{\Pi}_K\vec{v}$ coincide on the vertices and sides of K. Hence,

$$\Pi_K\vec{v} - \tilde{\Pi}_K\vec{v} = (\Pi_K\vec{v} - \tilde{\Pi}_K\vec{v})(a_{123})p_{123}.$$

Therefore,

$$(2.50) \qquad |\Pi_K\vec{v}-\tilde{\Pi}_K\vec{v}|_{1,K} = |p_{123}|_{1,K} \|(\Pi_K\vec{v}-\tilde{\Pi}_K\vec{v})(a_{123})\|,$$

where again $\|\cdot\|$ denotes the Euclidean norm of $\mathbb{R}^2$. By Lemma 2.1,

(2.51) $$|p_{123}|_{1,K} \leq C_2 \|B_K^{-1}\| \, |\det(B_K)|^{\frac{1}{2}}.$$

Next,

$$\int_K (\Pi_K \vec{v} - \tilde{\Pi}_K \vec{v}) dx = (\Pi_K \vec{v} - \tilde{\Pi}_K \vec{v})(a_{123}) \int_K p_{123} \, dx.$$

By (2.45') and (2.46) and since $\tilde{\Pi}_K \vec{v} = \Pi_K \vec{v}$ on ∂K, we get :

$$\int_K (\Pi_K \vec{v} - \tilde{\Pi}_K \vec{v})_\ell dx = \int_{\partial K} x_\ell \, \gamma_\nu (\tilde{\Pi}_K \vec{v} - \vec{v}) d\sigma \quad \text{for } \ell = 1,2.$$

Also

$$\left| \int_K p_{123} \, dx \right| = |\det(B_K)| \left| \int_{\hat{K}} \hat{p}_{123} \, d\hat{x} \right| \geq C_3 |\det(B_K)|.$$

Then, the last three statements yield :

(2.52) $$|(\Pi_K \vec{v} - \tilde{\Pi}_K \vec{v})_\ell (a_{123})| \leq C_4 |\det(B_K)|^{-1} \left| \int_{\partial K} x_\ell \, \gamma_\nu (\tilde{\Pi}_K \vec{v} - \vec{v}) d\sigma \right|.$$

Let us revert to the reference triangle in the above line integral. Let K' be a side of K. Without loss of generality, we can assume that K' and its corresponding image $\hat{K}'$ lie respectively on the x_1 - and $\hat{x}_1$ - axes. Then, the jacobian of the transformation F_K restricted to $\hat{K}'$ is $|\det(B_K')|$, where B_K' is obtained by deleting the first line and first column of B_K (thus B_K' is a scalar). In view of the second condition of (2.45), we have :

$$\int_{K'} x_\ell \, \gamma_\nu (\tilde{\Pi}_K \vec{v} - \vec{v}) d\sigma = |\det(B_K')| \int_{\hat{K}'} (B_K \hat{x})_\ell \, \gamma_{\hat{\nu}} (\tilde{\Pi}_{\hat{K}} \hat{\vec{v}} - \hat{\vec{v}}) d\hat{\sigma}.$$

Therefore

$$\left| \int_{K'} x_\ell \, \gamma_\nu (\tilde{\Pi}_K \vec{v} - \vec{v}) d\sigma \right| \leq |\det(B_K')| \, \|B_K\| \, \|\tilde{\Pi}_{\hat{K}} \hat{\vec{v}} - \hat{\vec{v}}\|_{0,\hat{K}'} .$$

Since P_2 is invariant under $\tilde{\Pi}_{\hat{K}}$,we can apply Corollary 2.2 :

$$\|\tilde{\Pi}_{\hat{K}} \hat{\vec{v}} - \hat{\vec{v}}\|_{0,\hat{K}'} \leq C_5 |\hat{\vec{v}}|_{k+1,\hat{K}} \quad \text{for } k = 1,2.$$

Hence

$$\left| \int_{K'} x_\ell \, \gamma_\nu (\tilde{\Pi}_K \vec{v} - \vec{v}) d\sigma \right| \leq C_6 |\det(B_K')| \, |\det(B_K)|^{-\frac{1}{2}} \|B_K\|^{k+2} |\vec{v}|_{k+1,K} .$$

By Cramer's rule, we easily derive that

$$|\det(B_K')| \leq |\det(B_K)| \, \|B_K^{-1}\| .$$

When substituted in (2.52), these two inequalities yield :

(2.53) $$|(\Pi_K\vec{v}-\tilde{\Pi}_K\vec{v})_\ell(a_{123})| \leq C_7|\det(B_K)|^{-\frac{1}{2}}\|B_K\|^{k+2}\|B_K^{-1}\|\,|\vec{v}|_{k+1,K} .$$

Then, (2.50), (2.51) and (2.53) give :

$$|\Pi_K\vec{v}-\tilde{\Pi}_K\vec{v}|_{1,K} \leq C_8\|B_K\|^{k+2}\|B_K^{-1}\|^2\|\vec{v}\|_{k+1,K} ;$$

hence, by (2.23) :

(2.54) $$|\Pi_K\vec{v}-\tilde{\Pi}_K\vec{v}|_{1,K} \leq C_9\sigma_K^2 h_K^k\|\vec{v}\|_{k+1,K} \quad \text{for } k=1 \text{ or } 2.$$

The required bound follows from (2.49) and (2.54). ■

<u>THEOREM 2.6</u>

<u>Suppose that the solution $(\vec{u},p)$ of the Stokes problem belongs to $[H^3(\Omega)]^2 \times (H^2(\Omega)\cap L_0^2(\Omega))$. Let the spaces W_h and M_h be defined by (2.44).</u>

1°/ <u>If the family $\mathcal{T}_h$ is regular, we have the following error bound :</u>

(2.55) $$|\vec{u}-\vec{u}_h|_{1,\Omega} \leq C_1h^2(\|\vec{u}\|_{3,\Omega}+\|p\|_{2,\Omega}) .$$

2°/ <u>If, in addition, Ω is convex, then (2.55) can be improved :</u>

$$\|\vec{u}-\vec{u}_h\|_{0,\Omega} \leq C_2h^3(\|\vec{u}\|_{3,\Omega}+\|p\|_{2,\Omega}).$$

3°/ <u>Moreover, if the family $\mathcal{T}_h$ is uniformly regular, then</u>

$$\|p-p_h\|_{0,\Omega} \leq C_3h^2(\|\vec{u}\|_{3,\Omega}+\|p\|_{2,\Omega}).$$

<u>Proof</u>

The first part is an immediate consequence of the choice of r_h and Lemma 2.7.

For the sake of brevity, we omit the proofs of the other two parts, as they are very similar to those derived for Theorem 2.5. ■

<u>Remark 2.2.</u>

The approximation proposed in this section can be easily extended to quadrilateral elements with the same order of accuracy. This method can also be applied to the three-dimensional case. ■

2.4. NUMERICAL SOLUTION BY REGULARIZATION

The methods we have studied lead to a large system of linear equations involving both $\vec{u}_h$ and p_h. Rather than solving this system directly (by means of a Gauss method, for instance), we can reduce it to smaller systems by separating the computation of $\vec{u}_h$ from that of p_h. This can be achieved by the regularization algorithm developed in Chapter I §4. Let us first apply it to the continuous Stokes problem.

We have the following situation :

$$X = [H_0^1(\Omega)]^n, \quad M = L_0^2(\Omega);$$

$$a(\vec{u},\vec{v}) = \nu(\overrightarrow{\text{grad}}\,\vec{u}, \overrightarrow{\text{grad}}\,\vec{v}),$$

$$b(\vec{v},p) = -(p, \text{div}\,\vec{v}),$$

$$c(p,q) = (p,q).$$

Recall that the statement of problem (Q) is :

$$(Q) \qquad \begin{cases} -\nu\Delta\vec{u} + \overrightarrow{\text{grad}}\,p = \vec{f} \\ \text{div}\,\vec{u} = 0 \end{cases} \text{in } \Omega, \qquad \vec{u}|_\Gamma = \vec{0}.$$

Then problem (Q^ε) reads as follows :

$$(Q^\varepsilon) \qquad \begin{cases} -\nu\Delta\vec{u}^\varepsilon + \overrightarrow{\text{grad}}\,p^\varepsilon = \vec{f} \\ \text{div}\,\vec{u}^\varepsilon = -\varepsilon p^\varepsilon \end{cases} \text{in } \Omega, \qquad \vec{u}^\varepsilon|_\Gamma = \vec{0}.$$

By eliminating p^ε, we get an equivalent second order elliptic problem with no constraint on $\vec{u}^\varepsilon$:

$$(P^\varepsilon) \qquad \begin{cases} -\nu\Delta\vec{u}^\varepsilon - \frac{1}{\varepsilon}\overrightarrow{\text{grad}}\,\text{div}\,\vec{u}^\varepsilon = \vec{f} \text{ in } \Omega, \\ \vec{u}^\varepsilon|_\Gamma = \vec{0}. \end{cases}$$

It is easy to verify all the hypotheses of Theorem 4.3 Chapter I ; therefore the following estimate holds :

$$(2.56) \qquad |\vec{u}-\vec{u}^\varepsilon|_{1,\Omega} + \| p + \frac{1}{\varepsilon}\text{div}\,\vec{u}^\varepsilon \|_{0,\Omega} \leqslant C\varepsilon \|\vec{f}\|_{-1,\Omega}.$$

Now, let us apply this technique to the general discrete problems of section 2.1. Problem (Q_h^ε) is :

$$(Q_h^\varepsilon)\begin{cases} \text{Find a pair } (\vec{u}_h^\varepsilon, p_h^\varepsilon) \in X_h \times M_h \text{ such that :} \\ (2.57) \qquad a(\vec{u}_h^\varepsilon, \vec{v}_h) - (p_h^\varepsilon, \operatorname{div} \vec{v}_h) = \langle \vec{f}, \vec{v}_h \rangle \quad \forall \vec{v}_h \in X_h, \\ (2.58) \qquad (\varepsilon p_h^\varepsilon + \operatorname{div} \vec{u}_h^\varepsilon, q_h) = 0 \quad \forall q_h \in M_h. \end{cases}$$

As usual, let ρ_h denote the orthogonal projection operator in $L^2(\Omega)$ onto M_h. Equation (2.58) may be written in the form :

$$p_h^\varepsilon = -\frac{1}{\varepsilon}\rho_h(\operatorname{div} \vec{u}_h^\varepsilon).$$

Therefore, we can eliminate p_h^ε from (2.57) and we get the equivalent analogue of (P^ε) :

$$(P_h^\varepsilon)\begin{cases} \text{Find } \vec{u}_h^\varepsilon \in X_h \text{ such that :} \\ a(\vec{u}_h^\varepsilon, \vec{v}_h) + \frac{1}{\varepsilon}\big(\rho_h(\operatorname{div} \vec{u}_h^\varepsilon),\ \rho_h(\operatorname{div} \vec{v}_h)\big) = \langle \vec{f}, \vec{v}_h \rangle \quad \forall \vec{v}_h \in X_h. \end{cases}$$

Of course problem (P_h^ε) offers a practical interest only if $\rho_h(\operatorname{div} \vec{v}_h)$ can be easily computed. But this is precisely the case of the examples of sections 2.2 and 2.3 because the functions of M_h are discontinuous and therefore ρ_h is a *local* operator, acting separately element by element.

Using again Theorem 4.3, Chapter I, one can easily check that :

$$(2.59) \qquad |\vec{u}_h - \vec{u}_h^\varepsilon|_{1,\Omega} + \| p_h + \frac{1}{\varepsilon}\rho_h(\operatorname{div} \vec{u}_h^\varepsilon) \|_{0,\Omega} \leq C\varepsilon \| \vec{f} \|_{-1,\Omega},$$

where the constant C is independent of ε and h.

<u>Remark 2.3.</u>

It is also possible to calculate $\vec{u}_h$ and p_h separately by the duality method of section 4.4, Chapter I. When applied to problem (Q), the duality algorithm gives :

$$(Q_m)\begin{cases} \text{Find a pair } (\vec{u}_{m+1}, \lambda_{m+1}) \in X \times M \text{ such that :} \\ a(\vec{u}_{m+1}, \vec{v}) + r(\operatorname{div} \vec{u}_{m+1}, \operatorname{div} \vec{v}) - (p_m, \operatorname{div} \vec{v}) = \langle \vec{f}, \vec{v} \rangle \quad \forall \vec{v} \in X, \\ (p_{m+1} - p_m, q) + \rho_m(\operatorname{div} \vec{u}_{m+1}, q) = 0 \quad \forall q \in M. \end{cases}$$

Here $\gamma = 1$ and $\alpha(r) = r$. For $r > 0$, Theorem 4.4, Chapter I asserts that if

$$(2.60) \qquad 0 < \inf_m \rho_m \leq \sup_m \rho_m < 2r$$

then

$$\lim_{m\to\infty} \{ \| \vec{u} - \vec{u}_m \|_X + \| \lambda - \lambda_m \|_M \} = 0.$$

When applied directly to problem (Q_h), the duality algorithm becomes :

$$(Q_h^m) \begin{cases} \text{Find a pair } (\vec{u}_h^{m+1}, p_h^{m+1}) \in X_h \times M_h, \text{ such that :} \\ a(\vec{u}_h^{m+1}, \vec{v}_h) + r(\rho_h(\operatorname{div} \vec{u}_h^{m+1}), \rho_h(\operatorname{div} \vec{v}_h)) - (p_h^m, \operatorname{div} \vec{v}_h) = \langle \vec{f}, \vec{v}_h \rangle \quad \forall \vec{v}_h \in X_h, \\ (p_h^{m+1} - p_h^m, q_h) + \rho_m(\operatorname{div} \vec{u}_h^{m+1}, q_h) = 0 \quad \forall q_h \in M_h. \end{cases}$$

Note that the first approximation p_h^0 may be chosen in $\mathcal{O}_h$ (i.e. need not be taken in M_h) and similarly p_0 can be taken in $L^2(\Omega)$.

It can be proved that u_h^m tends to $\vec{u}_h$ in $[H_0^1(\Omega)]^2$ and p_h^m tends to p_h in $L^2(\Omega)$ provided $r > 0$ and ρ_m satisfies (2.60). ∎

CHAPTER III

A MIXED FINITE ELEMENT METHOD FOR SOLVING THE STOKES PROBLEM

The preceding chapter brought out clearly the difficulties encountered in the practical construction of simple and continuous divergence-free velocity. Indeed, the examples we gave made use of continuous velocities with a small but non-vanishing divergence.

Here, instead, we shall develop a finite element method based upon discontinuous, but exactly divergence-free functions.

§ 1. MIXED APPROXIMATION OF AN ABSTRACT PROBLEM

In this paragraph, we concentrate again upon the abstract problem studied in Chapter I, §4, but we put it in a weaker setting leading to a mixed formulation. Then, we derive a mixed approximation from this formulation.

1.1. A MIXED VARIATIONAL PROBLEM

We put ourselves in the situation of section 4.1, Chapter I. Recall that problem (Q) is :

$$(Q)\begin{cases} \text{Find a pair } (u,\lambda) \text{ in } X\times M \text{ such that} & \\ (1.1) \quad a(u,v)+b(v,\lambda)=\langle \ell,v\rangle \quad \forall v\in X & \\ (1.2) \quad b(u,\mu)=\langle \chi,\mu\rangle \quad \forall \mu\in M, & \end{cases}$$

where, as usual, the bilinear forms a and b satisfy the hypotheses :

$$(1.3) \qquad a(v,v)\geqslant \alpha\|v\|_X^2,\ \alpha>0, \quad \forall v\in V$$

$$(1.4) \qquad \sup_{v\in X}\frac{b(v,\mu)}{\|v\|_X}\geqslant \beta\|\mu\|_M,\ \beta>0, \quad \forall \mu\in M.$$

These two hypotheses guarantee that the problem (Q) and its corresponding problem (P) are well posed:

$$(P)\begin{cases} \text{Find } u \text{ in } V(\chi) \text{ such that} \\ (1.5) \qquad a(u,v) = \langle \ell, v\rangle \quad \forall v \in V. \end{cases}$$

Now, let us give a weaker formulation of problem (Q). We introduce two Hilbert spaces $\tilde{X}$ and $\tilde{M}$, normed respectively by $\|\cdot\|_{\tilde{X}}$ and $\|\cdot\|_{\tilde{M}}$, such that

$$X \underset{d}{\hookrightarrow} \tilde{X}, \quad \tilde{M} \underset{d}{\hookrightarrow} M,$$

where the sign $\underset{d}{\hookrightarrow}$ means that the imbedding is dense and continuous.

Next, we consider two continuous bilinear forms

$$\tilde{a}(\cdot,\cdot) : \tilde{X} \times \tilde{X} \longmapsto \mathbb{R}, \quad \tilde{b}(\cdot,\cdot) : \tilde{X} \times \tilde{M} \longmapsto \mathbb{R}$$

and we set

$$\|\tilde{a}\| = \sup_{u,v \in \tilde{X}} \frac{\tilde{a}(u,v)}{\|u\|_{\tilde{X}} \|v\|_{\tilde{X}}}, \qquad \|\tilde{b}\| = \sup_{v \in \tilde{X}, \mu \in \tilde{M}} \frac{\tilde{b}(v,\mu)}{\|v\|_{\tilde{X}} \|\mu\|_{\tilde{M}}}.$$

These two bilinear forms are extensions of a and b in the sense that

$$(1.6) \qquad \tilde{a}(u,v) = a(u,v) \qquad \forall u,v \in X$$

$$(1.7) \qquad \tilde{b}(v,\mu) = b(v,\mu) \qquad \forall v \in X, \quad \forall \mu \in \tilde{M}.$$

In addition, we assume that ℓ, the right-hand side of (1.1), belongs to $\tilde{X}'$, the dual space of $\tilde{X}$, and we denote by $\langle\cdot,\cdot\rangle$ the duality between $\tilde{X}$ and $\tilde{X}'$. Then, we consider the following problem :

$$(\tilde{Q})\begin{cases} \text{Find a pair } (\tilde{u},\tilde{\lambda}) \in \tilde{X} \times \tilde{M} \text{ such that :} \\ (1.8) \qquad \tilde{a}(\tilde{u},v) + \tilde{b}(v,\tilde{\lambda}) = \langle \ell, v\rangle \quad \forall v \in \tilde{X} \\ (1.9) \qquad \tilde{b}(\tilde{u},\mu) = \langle \chi, \mu\rangle \quad \forall \mu \in \tilde{M}. \end{cases}$$

For each $\chi \in M'$, we define the affine variety :

$$(1.10) \qquad \tilde{V}(\chi) = \{v \in \tilde{X} \ ; \ \tilde{b}(v,\mu) = \langle \chi,\mu\rangle \quad \forall \mu \in \tilde{M}\},$$

and the following closed subspace of $\tilde{X}$:

$$(1.11) \qquad \tilde{V} = \tilde{V}(0) = \{v \in \tilde{X} \ ; \ \tilde{b}(v,\mu) = 0 \quad \forall \mu \in \tilde{M}\}.$$

Equality (1.7) implies that :

(1.12) $$V \subset \tilde{V}, \quad V(\chi) \subset \tilde{V}(\chi).$$

With $(\tilde{Q})$, we associate the following problem :

$$(\tilde{P}) \begin{cases} \text{Find } \tilde{u} \text{ in } \tilde{V}(\chi) \text{ such that} \\ \tilde{a}(\tilde{u},v) = \langle \ell, v \rangle \quad \forall v \in \tilde{V} \quad (1.13) \end{cases}$$

In order to analyze conveniently problems $(\tilde{P})$ and $(\tilde{Q})$, we make the following assumptions on the forms $\tilde{a}$ and $\tilde{b}$:

i) $\tilde{a}$ is $\tilde{V}$-elliptic ; i.e. there exists a constant $\tilde{\alpha} > 0$ such that :

(1.14) $$\tilde{a}(v,v) \geq \tilde{\alpha} \|v\|^2_{\tilde{X}} \quad \forall v \in \tilde{V} ;$$

ii) $\tilde{b}$ satisfies a weak inf-sup condition in the sense that there exists a constant $\tilde{\beta} > 0$ such that

(1.15) $$\inf_{\mu \in M} \sup_{v \in X} \frac{\tilde{b}(v,\mu)}{\|\mu\|_M \|v\|_{\tilde{X}}} \geq \tilde{\beta}.$$

This condition is indeed weaker than the ordinary inf-sup condition (4.9), Chapter I because it involves the norm of M instead of $\tilde{M}$. Strictly speaking, it is not sufficient to ensure that problem $(\tilde{Q})$ is well posed. The next theorem tackles this difficulty.

<u>THEOREM 1.1.</u>

<u>Let</u> (u,λ) <u>be the solution of problem</u> (Q).

1°/ <u>If $\tilde{a}$ satisfies (1.14), then problem $(\tilde{P})$ has exactly one solution $\tilde{u}$ in $\tilde{V}(\chi)$. Moreover, if $\tilde{u}$ also belongs to $V(\chi)$, or if V is dense in $\tilde{V}$, then $\tilde{u} = u$.</u>

2°/ <u>In addition, if $\tilde{b}$ satisfies (1.15) and if λ belongs to $\tilde{M}$ then the pair (u,λ) is the only solution of $(\tilde{Q})$.</u>

<u>Proof</u>

1°/ The ellipticity of $\tilde{a}$ and the condition (1.4) on b imply that $(\tilde{P})$ has one and only one solution $\tilde{u}$ in $\tilde{V}(\chi)$. If $\tilde{u} \in V(\chi)$, we see from (1.6) that $\tilde{u}$ is a solution of (P) ; hence $\tilde{u} = u$, since (P) has exactly one solution. Otherwise, we assume that V is dense in $\tilde{V}$; then (1.5) and (1.6) imply that u is a solution of $(\tilde{P})$. Therefore $u = \tilde{u}$.

2°/ In addition, suppose that $\lambda \in \widetilde{M}$. Then, by virtue of (1.6) and (1.7), (1.1) becomes :

$$\tilde{a}(u,v) + \tilde{b}(v,\lambda) = \langle \ell, v \rangle \quad \forall v \in X.$$

As X is dense in $\widetilde{X}$, this shows that (u,λ) is a solution of $(\widetilde{Q})$. Conversely, we must prove that $(\widetilde{Q})$ has a unique solution. Obviously, its first component is unique. Then, assume that

$$\tilde{b}(v,\lambda) = 0 \quad \forall v \in \widetilde{X}.$$

With hypothesis (1.15), this implies that $\lambda = 0$. ■

Remarks 1.1.

1°/ As far as this proof is concerned, we can replace condition (1.15) by the weaker statement : $\mathrm{Ker}\,\widetilde{B}' = \{0\}$, where $\widetilde{B}' \in \mathcal{L}(\widetilde{M};\widetilde{X}')$ is defined by :

$$\langle \widetilde{B}'\mu, v \rangle = \tilde{b}(v,\mu) \quad \forall \mu \in \widetilde{M}, \quad \forall v \in \widetilde{X}.$$

2°/ Further on, we shall study applications where the approximation of problems $(\widetilde{P})$ and $(\widetilde{Q})$ is simpler than that of problems (P) and (Q). ■

1.2. ABSTRACT MIXED APPROXIMATION

Throughout this section, we assume that the hypotheses of Theorem 1.1 are valid. For each h, let X_h and M_h be two finite-dimensional spaces satisfying:

$$X_h \subset \widetilde{X}, \quad M_h \subset \widetilde{M}. \tag{1.16}$$

We approximate problem $(\widetilde{Q})$ by :

(Q_h) Find a pair $(u_h,\lambda_h) \in X_h \times M_h$ such that

$$\tilde{a}(u_h,v_h) + \tilde{b}(v_h,\lambda_h) = \langle \ell, v_h \rangle \quad \forall v_h \in X_h, \tag{1.17}$$

$$\tilde{b}(u_h,\mu_h) = \langle \chi, \mu_h \rangle \quad \forall \mu_h \in M_h. \tag{1.18}$$

Again, we define

$$V_h(\chi) = \{v_h \in X_h \ ; \ \tilde{b}(v_h,\mu_h) = \langle \chi, \mu_h \rangle \quad \forall \mu_h \in M_h\} \tag{1.19}$$

and

$$V_h = V_h(0) = \{v_h \in X_h \ ; \ \tilde{b}(v_h,\mu_h) = 0 \quad \forall \mu_h \in M_h\}. \tag{1.20}$$

Next, with (Q_h) we associate the following problem :

$$(P_h)\begin{cases} \text{Find } u_h \in V_h(\chi) \text{ such that} \\ (1.21) \qquad \tilde{a}(u_h,v_h) = \langle \ell, v_h \rangle \quad \forall v_h \in V_h. \end{cases}$$

Here again, V_h is generally not included in $\tilde{V}$ and therefore (P_h) is an external approximation of $(\tilde{P})$.

In order to derive error estimates for u_h and λ_h, we make the following assumptions, analogous to (1.14) and (1.15) :

i) there exists a constant $\alpha^\star > 0$ such that

$$(1.22) \qquad \tilde{a}(v_h,v_h) \geq \alpha^\star \| v_h \|^2_{\tilde{X}} \quad \forall v_h \in V_h \ ;$$

ii) there exists a constant $\beta^\star > 0$ such that

$$(1.23) \qquad \sup_{v_h \in X_h} \frac{\tilde{b}(v_h,\mu_h)}{\| v_h \|_{\tilde{X}}} \geq \beta^\star \| \mu_h \|_M \quad \forall \mu_h \in M_h.$$

The next theorem is a natural extension of Theorem 1.1, Chapter II.

<u>THEOREM 1.2</u>

1°/ <u>Suppose</u> $V_h(\chi)$ <u>is not empty and</u> $\tilde{a}$ <u>satisfies (1.22). Then problem</u> (P_h) <u>has one and only one solution</u> $u_h \in V_h(\chi)$ <u>and the following error bound holds</u> :

$$(1.24) \qquad \| u-u_h \|_{\tilde{X}} \leq \left(1 + \frac{\|\tilde{a}\|}{\alpha^\star}\right) \inf_{v_h \in V_h(\chi)} \| u-v_h \|_{\tilde{X}} + \frac{1}{\alpha^\star} \inf_{\mu_h \in M_h} \sup_{v_h \in V_h} \frac{|\tilde{b}(v_h,\lambda-\mu_h)|}{\| v_h \|_{\tilde{X}}} .$$

2°/ <u>Suppose</u> $\tilde{b}$ <u>satisfies (1.23). Then</u> $V_h(\chi)$ <u>is not empty and problem</u> (Q_h) <u>has exactly one solution</u> (u_h,λ_h), <u>where</u> u_h <u>is the solution of</u> (P_h). <u>Furthermore,</u> λ_h <u>satisfies the error estimate</u> :

$$(1.25) \qquad \| \lambda-\lambda_h \|_M \leq \frac{\|\tilde{a}\|}{\beta^\star} \| u-u_h \|_{\tilde{X}} + \inf_{\mu_h \in M_h} \left\{ \frac{\|\tilde{b}\|}{\beta^\star} \| \lambda-\mu_h \|_{\tilde{M}} + \| \lambda-\mu_h \|_M \right\}.$$

<u>Proof</u>

1°/ The idea of the proof is similar to that of Theorem 1.1, Chapter II. The existence and uniqueness of the solution u_h of (P_h) follows from (1.22) and

Lax-Milgram's theorem, provided $V_h(\chi)$ is not empty.

Now, let w_h be any element of $V_h(\chi)$ and let $v_h = u_h - w_h \in V_h$. Then

$$\tilde{a}(v_h,v_h) = \langle \ell, v_h \rangle_{\tilde{X}} - \tilde{a}(w_h,v_h),$$

that is

$$\tilde{a}(v_h,v_h) = \tilde{a}(u-w_h,v_h) + \tilde{b}(v_h,\lambda-\mu_h) \tag{1.26}$$

$$\forall \mu_h \in M_h, \quad \forall w_h \in V_h(\chi).$$

From (1.22) we derive :

$$\alpha^* \| v_h \|_{\tilde{X}} \leq \| \tilde{a} \| \, \| u-w_h \|_{\tilde{X}} + \sup_{v_h \in V_h} \left\{ \frac{|\tilde{b}(v_h,\lambda-\mu_h)|}{\| v_h \|_{\tilde{X}}} \right\},$$

which gives immediately (1.24).

2°/ Since M_h is finite-dimensional, the condition (1.23) implies the classical inf-sup condition on M_h, eventually with a constant that depends upon h. Therefore $V_h(\chi)$ is not empty and (Q_h) has exactly one solution (u_h,λ_h), where u_h satisfies (P_h). Moreover, the following equation holds for any v_h in X_h and μ_h in M_h :

$$\tilde{b}(v_h,\lambda_h-\mu_h) = \tilde{a}(u-u_h,v_h) + \tilde{b}(v_h,\lambda-\mu_h).$$

Then (1.23) implies that

$$\beta^* \| \lambda_h-\mu_h \|_M \leq \| \tilde{a} \| \, \| u-u_h \|_{\tilde{X}} + \| \tilde{b} \| \, \| \lambda-\mu_h \|_{\tilde{M}},$$

and (1.25) is established. ∎

Note that the estimate (1.25) is not optimal since it gives an upper bound for $\| \lambda-\lambda_h \|_M$ in terms of $\| \lambda-\mu_h \|_{\tilde{M}}$ whereas, in general, $\tilde{M}$ is strictly included in M.

We also remark that it is usually difficult to evaluate directly an expression like $\inf_{v_h \in V_h(\chi)} \| u-v_h \|_{\tilde{X}}$. In fact, it is possible to reduce this expression to the approximation error in X_h, although the process is not always optimal. As the dimension of M_h is finite, there exists a constant S(h) such that

$$\| \mu_h \|_{\tilde{M}} \leq S(h) \| \mu_h \|_M \quad \forall \mu_h \in M_h. \tag{1.27}$$

With this, we prove the following result :

COROLLARY 1.1.

Under hypotheses (1.22) and (1.23), problem (Q_h) has a unique solution (u_h,λ_h) and there exists a constant C, depending solely upon α^*, β^*, $\|\tilde{a}\|$ and $\|\tilde{b}\|$, such that :

$$\|u-u_h\|_{\tilde{X}} + \|\lambda-\lambda_h\|_M \leq C\{\big(1+S(h)\big) \inf_{v_h\in X_h} \|u-v_h\|_{\tilde{X}} + \inf_{\mu_h\in M_h} \|\lambda-\mu_h\|_{\tilde{M}}\}. \tag{1.28}$$

Proof

The estimate (1.28) is an immediate consequence of (1.24) and (1.25), provided there exists a constant C such that

$$\inf_{w_h\in V_h(\chi)} \|u-w_h\|_{\tilde{X}} \leq C\big(1+S(h)\big) \inf_{v_h\in X_h} \|u-v_h\|_{\tilde{X}}. \tag{1.29}$$

Let us establish (1.29). By virtue of (1.27), condition (1.23) becomes :

$$\sup_{v_h\in X_h} \frac{\tilde{b}(v_h,\mu_h)}{\|v_h\|_{\tilde{X}}} \geq \beta^*\big(S(h)\big)^{-1} \|\mu_h\|_{\tilde{M}} \quad \forall \mu_h\in M_h. \tag{1.30}$$

Therefore, the statement of Lemma 4.1, Chapter I is valid with β replaced by $\beta^*\big(S(h)\big)^{-1}$ and we can proceed exactly like in the proof of (1.12), Chapter II. Thus, we obtain :

$$\|u-w_h\|_{\tilde{X}} \leq \Big(1+S(h)\frac{\|\tilde{b}\|}{\beta^*}\Big)\|u-v_h\|_{\tilde{X}} \quad \forall w_h\in V_h(\chi), \ \forall v_h\in X_h.$$

This proves (1.29). ■

We shall see in the examples that S(h) usually depends upon the dimension of M_h - and, more precisely, that S(h) tends to infinity as the dimension of M_h tends to infinity.

§ 2. APPLICATION TO THE HOMOGENEOUS STOKES PROBLEM

Here, we use the theory developed in the preceding paragraph to formulate and approximate the Stokes problem. For the sake of simplicity, we restrict ourselves to the two-dimensional case.

2.1. A MIXED FORMULATION OF STOKES EQUATIONS

Let Ω be a bounded domain of $\mathbb{R}^2$ with a Lipschitz continuous boundary Γ whose components are denoted by Γ_i, for $0 \leqslant i \leqslant p$, (cf. Figure 1). For $\vec{f}$ given in $[H^{-1}(\Omega)]^2$, the homogeneous Stokes equations in Ω are :

$$(2.1)\quad \left\{\begin{array}{l} \text{Find } (\vec{u},p) \text{ in } [H^1(\Omega)]^2 \times L_0^2(\Omega) \text{ satisfying} \\ \left.\begin{array}{r} -\nu\Delta\vec{u} + \overrightarrow{\text{grad}}\, p = \vec{f} \\ \text{div}\,\vec{u} = 0 \end{array}\right\} \text{ in } \Omega, \\ \qquad\qquad \vec{u}|_{\Gamma} = \vec{0}. \end{array}\right.$$

As usual, we set $V = \{\vec{v} \in [H_0^1(\Omega)]^2 \ ;\ \text{div}\,\vec{v} = 0\}$, and its corresponding space of of stream functions :

$$\Phi = \{\phi \in H^2(\Omega)\ ;\ \phi|_{\Gamma_0} = 0,$$
$$\phi|_{\Gamma_i} = \text{an arbitrary constant } c_i,\ 1 \leqslant i \leqslant p,\ \left.\frac{\partial\phi}{\partial\nu}\right|_{\Gamma} = 0\}.$$

We have seen in §5 Chapter I that problem (2.1) is equivalent to the following two problems :

$$(2.2)\quad \left\{\begin{array}{l} \text{Find } \vec{u} \in V \text{ such that} \\ \nu(\overrightarrow{\text{grad}}\,\vec{u}, \overrightarrow{\text{grad}}\,\vec{v}) = (\vec{f},\vec{v}) \quad \forall \vec{v} \in V\ ; \end{array}\right.$$

$$(2.3)\quad \left\{\begin{array}{l} \text{Find } \psi \in \Phi \text{ such that} \\ \nu(\Delta\psi,\Delta\phi) = (\vec{f}, \overrightarrow{\text{curl}}\,\phi) \quad \forall\phi \in \Phi. \end{array}\right.$$

As (2.2) and (2.3) are equivalent, we can use either one according to convenience.

Now, it has been proved (cf. Theorem 5.4, Chapter I) that

$$(\overrightarrow{\text{grad}}\,\vec{u}, \overrightarrow{\text{grad}}\,\vec{v}) = (\text{curl}\,\vec{u}, \text{curl}\,\vec{v}) \text{ for } \vec{u} \text{ (or } \vec{v}) \text{ in } [H^1(\Omega)]^2$$

and $\vec{v}$ (or $\vec{u}$) in V. Therefore, another equivalent formulation for (2.1) is :

$$(2.4)\quad \begin{cases} \text{Find } \vec{u} \text{ in } V \text{ satisfying :} \\ \nu(\operatorname{curl}\vec{u},\operatorname{curl}\vec{v}) = (\vec{f},\vec{v}) \quad \forall \vec{v}\in V . \end{cases}$$

Problem (2.4) lends itself readily to a useful mixed formulation. First, we remark that another description of space V is :

$$(2.5)\qquad V_1 = \{(\vec{v},\theta) \in [H_0^1(\Omega)]^2 \times L^2(\Omega) \;;\; \operatorname{div}\vec{v} = 0,\ \theta = \operatorname{curl}\vec{v}\}.$$

Then problem (2.4) is equivalent to :

$$(P_1)\quad \begin{cases} \text{Find } (\vec{u},\omega)\in V_1 \text{ satisfying} \\ (2.6)\qquad \nu(\omega,\theta) = (\vec{f},\vec{v}) \quad \forall(\vec{v},\theta)\in V_1 . \end{cases}$$

We can also describe space V in terms of stream functions, since there is a one-to-one correspondence between Φ and V by the relation : $-\Delta\phi = \operatorname{curl}\vec{v}$. Thus, we can write :

$$(2.7)\qquad V_2 = \{(\phi,\theta)\in\Phi\times L^2(\Omega) \;;\; \theta = -\Delta\phi\}.$$

Then problem (2.3) is equivalent to :

$$(P_2)\quad \begin{cases} \text{Find } (\psi,\omega)\in V_2 \text{ such that} \\ (2.8)\qquad \nu(\omega,\theta) = (\vec{f},\overrightarrow{\operatorname{curl}}\,\phi) \quad \forall(\phi,\theta)\in V_2 . \end{cases}$$

Thus, we have introduced in these two problems $\omega = \operatorname{curl}\vec{u} = -\Delta\psi$ as an additional unknown. But a look at (2.7) shows that problems (P_1) and (P_2) are not satisfactory because their internal approximation requires the construction of finite-dimensional subspace of $H^2(\Omega)$; in practice, this is far from desirable. In order to avoid this difficulty, we shall introduce a Lagrange multiplier corresponding to the constraint $\omega = \operatorname{curl}\vec{u}$ (or equivalently $\omega = -\Delta\psi$) and then relax the regularity of the test functions, while retaining their divergence-free property, by following the pattern of §1.

Consider first problem (P_1) - and, as there is no confusion, let us drop the subscript 1. This is precisely the type of problem studied in §1, with the following choice of spaces and bilinear forms :

$$X = \{\vec{v}\in[H_0^1(\Omega)]^2 \;;\; \operatorname{div}\vec{v} = 0\}\times L^2(\Omega),\qquad M = L^2(\Omega),$$

$$a(u,v) = \nu(\omega,\theta) \quad \text{for } u = (\vec{u},\omega),\ v = (\vec{v},\theta) \in X,$$

and

$$b(v,\mu) = (\text{curl}\,\vec{v}-\theta,\mu) \text{ for } v = (\vec{v},\theta) \in X, \quad \mu \in M.$$

Then the space V defined by (2.5) is $\{v \in X \ ; \ b(v,\mu) = 0 \quad \forall \mu \in M\}$ and problem (Q) associated with (P) is :

Find a pair $(u,\lambda) \in X \times M$ satisfying

$$(Q) \begin{cases} (2.9) & a(u,v) + b(v,\lambda) = (\vec{f},\vec{v}) \quad \forall v \in X, \\ (2.10) & b(u,\mu) = 0 \quad \forall \mu \in M. \end{cases}$$

<u>THEOREM 2.1.</u>

<u>Problem (Q) has exactly one solution</u> $(u,\lambda) \in X \times M$ <u>and</u>

$$(2.11) \qquad u = (\vec{u},\text{curl}\,\vec{u}), \quad \lambda = \nu(\text{curl}\,\vec{u})$$

<u>where</u> $\vec{u}$ <u>is the solution of Stokes problem (2.1).</u>

<u>Proof.</u>

Since problem (P) is equivalent to Stokes problem, it suffices to verify the inf-sup condition on b in order to obtain the existence and uniqueness of λ. Thus, let $\mu \in M$ and let us pick $v = (\vec{0},-\mu)$ in X. Then

$$\sup_{v \in X} \frac{b(v,\mu)}{\|v\|_X} \geq \frac{\|\mu\|^2_{0,\Omega}}{\|\mu\|_{0,\Omega}} = \|\mu\|_M \quad ,$$

hence the inf-sup condition holds with $\beta = 1$.

It remains to show that $\lambda = \nu\omega$. Indeed, let $v = (\vec{v},\theta) \in X$ and let $u = (\vec{u},\omega) \in V$ be the solution of (P). Then

$$a(u,v) + b(v,\nu\omega) = \nu(\omega,\theta) + \nu(\text{curl}\,\vec{v}-\theta,\omega) = \nu(\text{curl}\,\vec{v},\omega)$$
$$= \nu(\text{curl}\,\vec{v},\ \text{curl}\,\vec{u}) = (\vec{f},\vec{v}).$$

As λ is unique, it follows that necessarily $\lambda = \nu\omega$. ∎

Now, we relax the regularity of our functions. We take :

$$\tilde{X} = \{\vec{v} \in H_0(\text{div};\Omega) \ ; \ \text{div}\,\vec{v} = 0\} \times L^2(\Omega), \quad \tilde{M} = H^1(\Omega),$$

$$\tilde{a}(u,v) = a(u,v) = \nu(\omega,\theta) \text{ for } u = (\vec{u},\omega),\ v = (\vec{v},\theta) \in \tilde{X},$$

and

$$\tilde{b}(v,\mu) = (\vec{v}, \overrightarrow{\text{curl}}\,\mu) - (\theta,\mu) \text{ for } v = (\vec{v},\theta) \in \tilde{X},\ \mu \in \tilde{M}.$$

Then

$$\tilde{V} = \{(\vec{v},\theta) \in \tilde{X}\ ;\ (\vec{v}, \overrightarrow{\text{curl}}\,\mu) = (\theta,\mu) \quad \forall \mu \in H^1(\Omega)\}.$$

Next, we recall the space $\tilde{\Phi}$ introduced in §3, Chapter I :

$$\tilde{\Phi} = \{\phi \in H^1(\Omega)\ ;\ \phi|_{\Gamma_0} = 0,$$
$$\phi|_{\Gamma_i} = \text{an arbitrary constant } c_i,\ 1 \leq i \leq p\},$$

and the one-to-one correspondence that was established between $\tilde{\Phi}$ and $\{\vec{v} \in H_0(\text{div}\ ;\ \Omega)\ ;\ \text{div}\ \vec{v} = 0\}$ by the relation

$$\vec{v} = \overrightarrow{\text{curl}}\,\phi\ .$$

Therefore $\tilde{V}$ can also be written equivalently as :

$$\tilde{V} = \{(\overrightarrow{\text{curl}}\,\phi,\theta)\ ;\ \phi \in \tilde{\Phi},\ \theta \in L^2(\Omega) \text{ with } (\overrightarrow{\text{curl}}\,\phi, \overrightarrow{\text{curl}}\,\mu) = (\theta,\mu) \quad \forall \mu \in H^1(\Omega)\}. \tag{2.12}$$

The problems $(\tilde{P})$ and $(\tilde{Q})$ associated with these spaces and norms are :

$(\tilde{Q})$ Find $u = (\vec{u},\omega) \in \tilde{X}$ and $\lambda \in H^1(\Omega)$ satisfying :

$$\nu(\omega,\theta) + (\vec{v}, \overrightarrow{\text{curl}}\,\lambda) - (\theta,\lambda) = (\vec{f},\vec{v}) \quad \forall v = (\vec{v},\theta) \in \tilde{X} \tag{2.13}$$

$$(\vec{u}, \overrightarrow{\text{curl}}\,\mu) - (\omega,\mu) = 0 \quad \forall \mu \in H^1(\Omega)\ ; \tag{2.14}$$

$(\tilde{P})$ Find $u = (\vec{u},\omega) \in \tilde{V}$ such that

$$\nu(\omega,\theta) = (\vec{f},\vec{v}) \quad \forall v = (\vec{v},\theta) \in \tilde{V}. \tag{2.15}$$

Let us check the hypotheses of § 1. It is well known that $\tilde{M} \underset{d}{\hookrightarrow} M$, and, as stated in § 2, Chapter I, $X \underset{d}{\hookrightarrow} \tilde{X}$. Equality (1.6) is obvious and (1.7) follows immediately from Green's formula. It remains to establish the ellipticity of $\tilde{a}$ and the inf-sup condition on $\tilde{b}$. As far as the ellipticity is concerned, we have :

$$\tilde{a}(v,v) = \nu \|\theta\|^2_{0,\Omega} \quad \text{for} \quad v = (\vec{v},\theta) \in \tilde{V}\ .$$

But by taking $\mu = \phi$ in the expression (2.12) of $\tilde{V}$ and applying Poincaré's inequality (Lemma 3.1, Chapter I) we get :

$$\|\vec{v}\|_{0,\Omega} = \|\overrightarrow{\text{curl}}\,\phi\|_{0,\Omega} \leq C\|\theta\|_{0,\Omega}\ .$$

Hence the mapping $v \longmapsto \|\theta\|_{0,\Omega}$ is a norm on $\tilde{V}$ equivalent to $\|v\|_{\tilde{X}}$ and therefore

$\tilde{a}$ is elliptic on $\tilde{V}$. Next, we turn to the inf-sup condition. For μ in $\tilde{M}$, we take $v = (\vec{0}, -\mu)$ in $\tilde{X}$; this yields directly the inequality (1.15) with $\tilde{\beta} = 1$.

In addition, we have the following result :

LEMMA 2.1.

The spaces V and $\tilde{V}$ are the same, i.e.

$$V = \tilde{V}.$$

Proof

Let $v = (\vec{v}, \theta) \in \tilde{V}$ and let us extend $\vec{v}$ and θ by zero outside Ω, i.e., we set

$$\tilde{\vec{v}} = \begin{cases} \vec{v} \text{ in } \Omega \\ \vec{0} \text{ in } \mathbb{R}^2 - \Omega \end{cases}, \quad \tilde{\theta} = \begin{cases} \theta \text{ in } \Omega \\ 0 \text{ in } \mathbb{R}^2 - \Omega \end{cases}, \quad \tilde{v} = (\tilde{\vec{v}}, \tilde{\theta}).$$

Thus $\tilde{\theta} \in L^2(\mathbb{R}^2)$ and $\tilde{\vec{v}} \in H(\mathrm{div}; \mathbb{R}^2)$ with $\mathrm{div}\, \tilde{\vec{v}} = 0$. Furthermore,

$$\int_{\mathbb{R}^2} (\tilde{\vec{v}} \cdot \overrightarrow{\mathrm{curl}}\, \mu - \tilde{\theta}\mu)\, dx = 0 \quad \forall \mu \in H^1(\mathbb{R}^2).$$

As a consequence,

$$\tilde{\theta} = \mathrm{curl}\, \tilde{\vec{v}} \text{ in } \mathbb{R}^2 .$$

It is easy to see, by means of Fourier transforms, that the above statements imply that $\tilde{\vec{v}} \in [H^1(\mathbb{R}^2)]^2$ and hence $\vec{v} \in (H^1(\Omega))^2$.

Now, on one hand $\vec{v} \cdot \vec{\nu} = 0$ on Γ and on the order hand

$$0 = (\vec{v}, \overrightarrow{\mathrm{curl}}\, \mu) - (\theta, \mu) = (\mathrm{curl}\, \vec{v} - \theta, \mu) + \int_\Gamma \vec{v} \cdot \vec{\tau}\, \mu\, d\sigma.$$

Therefore $\vec{v} \cdot \vec{\tau} = 0$ on Γ. Hence $\vec{v} \in (H_0^1(\Omega))^2$ and $\tilde{V} \subset V$, thus proving the equality. ∎

From Theorem 1.1 and the above results, we infer the next theorem.

THEOREM 2.2.

Problem $(\tilde{P})$ has a unique solution $u = (\vec{u}, \mathrm{curl}\, \vec{u}) \in V$, where $\vec{u}$ is the solution of the Stokes problem. Moreover, if $\mathrm{curl}\, \vec{u} \in H^1(\Omega)$, problems (Q) and $(\tilde{Q})$ are equivalent.

Of course, every statement above has its equivalent counterpart in terms of stream functions. Thus, by setting $X = \Phi \times L^2(\Omega)$, we get problem (Q_2) associated with (P_2) (again, we drop the subscript) :

$$(Q)\begin{cases} \text{Find } (\psi,\omega)\in X \text{ and } \lambda\in L^2(\Omega) \text{ satisfying :} \\ (2.16) \qquad \nu(\omega,\theta) - (\Delta\phi+\theta,\lambda) = (\vec{f},\overrightarrow{\text{curl}}\,\phi) \quad \forall(\phi,\theta)\in X, \\ (2.17) \qquad (\Delta\psi+\omega,\mu) = 0 \quad \forall\mu\in L^2(\Omega). \end{cases}$$

Then Theorem 2.1 implies that problem (Q) has the only solution : $\big((\psi,\omega = -\Delta\psi),\ \lambda = \nu\omega\big)$, where ψ is the stream function of $\vec{u}$ in Φ, and, as usual, $\vec{u}$ is the solution of the Stokes problem.

Similarly, we relax the regularity of (Q) by setting

$$(2.18) \qquad \tilde{X} = \tilde{\Phi}\times L^2(\Omega), \quad \tilde{M} = H^1(\Omega).$$

We keep the form $\tilde{a}$ unchanged and we express $\tilde{b}$ by :

$$\tilde{b}(v,\mu) = (\overrightarrow{\text{curl}}\,\phi,\overrightarrow{\text{curl}}\,\mu) - (\theta,\mu) \quad \forall v = (\phi,\theta)\in\tilde{X}, \quad \mu\in\tilde{M}.$$

The weak statement of (Q) is :

$$(\tilde{Q})\begin{cases} \text{Find } (\psi,\omega)\in\tilde{X} \text{ and } \lambda\in\tilde{M} \text{ such that} \\ (2.19) \quad \nu(\omega,\theta) + (\overrightarrow{\text{curl}}\,\phi,\overrightarrow{\text{curl}}\,\lambda) - (\theta,\lambda) = (\vec{f},\overrightarrow{\text{curl}}\,\phi) \quad \forall(\phi,\theta)\in\tilde{X}, \\ (2.20) \qquad (\overrightarrow{\text{curl}}\,\psi,\overrightarrow{\text{curl}}\,\mu) - (\omega,\mu) = 0 \quad \forall\mu\in\tilde{M}. \end{cases}$$

Likewise, if $\Delta\psi\in H^1(\Omega)$, Theorem 2.2. asserts that problem $(\tilde{Q})$ is equivalent to problem (Q).

Throughout the remainder of this paragraph, we shall suppose that the conclusions of Theorem 2.2. hold ; this will enable us to work indifferently either with stream functions or with velocities.

2.2. A MIXED METHOD FOR STOKES PROBLEM

Let us adapt to problem $(\tilde{Q})$ the general approximation developed in section 1.2. We introduce three finite-dimensional spaces : Φ_h, Θ_h and M_h such that

$$(2.21) \qquad \Phi_h\subset\tilde{\Phi}, \quad \Theta_h\subset L^2(\Omega), \quad M_h\subset H^1(\Omega).$$

Then we set

$$(2.22) \qquad X_h = \Phi_h\times\Theta_h,$$

and we define V_h by (1.20).

According to the definition (2.18), $X_h \subset \widetilde{X}$ and $M_h \subset \widetilde{M}$. The next two lemmas will deal with the conditions (1.22) and (1.23).

LEMMA 2.2.

If $\Phi_h \subset M_h$ there exist two constants $C>0$ and $\alpha^\star>0$, independent of h, such that :

$$\left.\begin{aligned} &(2.23) && |\phi_h|_{1,\Omega} \leqslant C\|\theta_h\|_{0,\Omega} \\ &(2.24) && \nu\|\theta_h\|^2_{0,\Omega} \geqslant \alpha^\star \|v_h\|^2_{\widetilde{X}} \end{aligned}\right\} \forall v_h = (\phi_h,\theta_h) \in V_h.$$

Proof.

By definition, a function $v_h = (\phi_h,\theta_h)$ in V_h satisfies :

$$(\overrightarrow{\mathrm{curl}}\,\phi_h, \overrightarrow{\mathrm{curl}}\,\mu_h) = (\theta_h,\mu_h) \quad \forall \mu_h \in M_h.$$

As $\Phi_h \subset M_h$, we can take $\mu_h = \phi_h$ in this equality. It yields :

$$|\phi_h|^2_{1,\Omega} \leqslant \|\theta_h\|_{0,\Omega}\|\phi_h\|_{0,\Omega}.$$

But, since $\Phi_h \subset \widetilde{\Phi}$, we can apply Lemma 3.1, Chapter I, to ϕ_h :

$$\|\phi_h\|_{0,\Omega} \leqslant C|\phi_h|_{1,\Omega},$$

thus proving (2.23). As a consequence, the mapping $v_h \mapsto \|\theta_h\|_{0,\Omega}$ is a norm on V_h uniformly equivalent to $\|v_h\|_{\widetilde{X}}$. In particular (2.24) holds and therefore the form $\widetilde{a}$ is uniformly elliptic on V_h. ■

LEMMA 2.3.

If $M_h \subset \Theta_h$ then

$$(2.25) \qquad \sup_{v_h \in X_h} \frac{\widetilde{b}(v_h,\mu_h)}{\|v_h\|_{\widetilde{X}}} \geqslant \|\mu_h\|_{0,\Omega} \quad \forall \mu_h \in M_h.$$

Proof

Let $\mu_h \in M_h$. As $M_h \subset \Theta_h$, the pair $v_h = (\vec{0},-\mu_h)$ belongs to X_h and for this v_h, we have :

$$\frac{\widetilde{b}(v_h,\mu_h)}{\|v_h\|_{\widetilde{X}}} = \|\mu_h\|_{0,\Omega}.$$ ■

THEOREM 2.3.

1°/ Suppose that

$$\Phi_h \subset M_h \subset \Theta_h \quad . \tag{2.26}$$

Then problem (Q_h) has exactly one solution $\left(u_h = (\psi_h,\omega_h),\lambda_h\right) \in X_h \times M_h$.

2°/ Moreover, if $M_h = \Theta_h$ then $\lambda_h = \nu\omega_h$.

Proof.

Part 1 follows immediately from Lemmas 2.2 and 2.3 and Theorem 1.2.

Let us show that $\lambda_h = \nu\omega_h$. First we remark that when $\Theta_h = M_h$, each function ϕ_h of Φ_h has a unique function θ_h in Θ_h such that the pair (ϕ_h,θ_h) belongs to V_h. Next, as $\Theta_h = M_h \subset H^1(\Omega)$, we have :

$$\tilde{a}(u_h,v_h) + \tilde{b}(v_h,\nu\omega_h) = \nu(\omega_h,\theta_h) + \nu(\overrightarrow{\mathrm{curl}}\,\phi_h,\overrightarrow{\mathrm{curl}}\,\omega_h) - \nu(\theta_h,\omega_h).$$

Hence, we must establish that

$$\nu(\overrightarrow{\mathrm{curl}}\,\phi_h,\overrightarrow{\mathrm{curl}}\,\omega_h) = (\vec{f},\overrightarrow{\mathrm{curl}}\,\phi_h) \qquad \forall\phi_h \in \Phi_h. \tag{2.27}$$

But, v_h satisfies $(\overrightarrow{\mathrm{curl}}\,\phi_h,\overrightarrow{\mathrm{curl}}\,\mu_h) = (\theta_h,\mu_h) \quad \forall\mu_h \in M_h$.

Here we can take $\mu_h = \omega_h$:

$$(\overrightarrow{\mathrm{curl}}\,\phi_h,\overrightarrow{\mathrm{curl}}\,\omega_h) = (\theta_h,\omega_h). \tag{2.28}$$

Finally, since $u_h = (\psi_h,\omega_h)$ is the solution of (P_h), we have :

$$\nu(\omega_h,\theta_h) = (\vec{f},\overrightarrow{\mathrm{curl}}\,\phi_h) \qquad \forall v_h = (\phi_h,\theta_h) \in V_h. \tag{2.29}$$

Therefore (2.27) follows from (2.28) and (2.29). ■

Remark 2.1.

When $\Theta_h = M_h$, we can eliminate entirely the Lagrange multiplier λ_h from problem (Q_h). The equations for (ψ_h,ω_h) become :

$$(2.30)\left\{\begin{array}{l} \nu(\overrightarrow{\mathrm{curl}}\,\omega_h,\overrightarrow{\mathrm{curl}}\,\phi_h) = (\vec{f},\overrightarrow{\mathrm{curl}}\,\phi_h) \qquad \forall\phi_h \in \Phi_h \\ \text{and} \\ (\overrightarrow{\mathrm{curl}}\,\psi_h,\overrightarrow{\mathrm{curl}}\,\mu_h) = (\omega_h,\mu_h) \qquad \forall\mu_h \in \Theta_h. \end{array}\right.$$

Note that the equations (2.30) are in fact a straightforward discretization of the problem :

$$(2.31)\quad \begin{cases} \nu(\overrightarrow{\mathrm{curl}}\,\omega,\overrightarrow{\mathrm{curl}}\,\phi) = (\vec{f},\overrightarrow{\mathrm{curl}}\,\phi) \quad \forall\phi\in\tilde{\Phi} \\ \text{and} \\ (\overrightarrow{\mathrm{curl}}\,\psi,\overrightarrow{\mathrm{curl}}\,\mu) = (\omega,\mu) \quad \forall\mu\in H^1(\Omega). \end{cases}$$ ∎

From Corollary 1.1 and Theorem 2.3, we derive the next corollary :

COROLLARY 2.1.

If $\Phi_h \subset M_h = \Theta_h$ there exists a constant $C>0$, independent of h, such that :

$$(2.32)\quad \begin{aligned} &|\psi-\psi_h|_{1,\Omega} + \|\omega-\omega_h\|_{0,\Omega} + \|\nu\omega-\lambda_h\|_{0,\Omega} \\ &\leq C\Big[(1+S(h))\{\inf_{\phi_h\in\Phi_h}|\psi-\phi_h|_{1,\Omega} + \inf_{\theta_h\in\Theta_h}\|\omega-\theta_h\|_{0,\Omega}\} \\ &\quad + \inf_{\mu_h\in M_h}\|\nu\omega-\mu_h\|_{1,\Omega}\Big]. \end{aligned}$$ ∎

However, in order to derive a more precise error estimate, it is necessary to specify the choice of the spaces Φ_h, Θ_h and M_h and thereby evaluate the quantity $S(h)$.

2.3. APPLICATION TO FINITE ELEMENTS OF DEGREE ℓ

To simplify the discussion, we assume in this section that Ω is a polygonal domain of $\mathbb{R}^2$. Let $\mathcal{T}_h$ be a family of triangulations of $\bar{\Omega}$ consisting of triangles whose diameters are bounded by h. We suppose that the family $\mathcal{T}_h$ is uniformly regular as h tends to zero (cf. definition 2.3, Chapter II), namely

$$\tau h \leq h_K \leq \sigma\rho_K \quad \forall K\in\mathcal{T}_h.$$

For a given integer $\ell \geq 1$, we choose the following finite element spaces :

$$(2.33)\quad \Theta_h = M_h = \{\theta_h \in \mathcal{C}^0(\bar{\Omega})\ ;\ \theta_h|_K \in P_\ell \quad \forall K\in\mathcal{T}_h\},$$

$$(2.34)\quad \begin{aligned} \Phi_h = \Theta_h \cap \tilde{\Phi} = \{\phi_h \in \mathcal{C}^0(\bar{\Omega})\ ;\ &\phi_h|_K \in P_\ell \quad \forall K\in\mathcal{T}_h,\ \phi_h|_{\Gamma_0} = 0, \\ &\phi_h|_{\Gamma_i} = \text{an arbitrary constant } c_i,\ 1\leq i\leq p\}. \end{aligned}$$

The following lemma establishes a bound for $S(h)$.

LEMMA 2.4.

Let $\mathcal{T}_h$ be a uniformly regular family of triangulations of $\bar{\Omega}$ and let the space M_h be defined by (2.33). Then there exists a constant A, independent of h, <u>such that</u>

(2.35) $$S(h) \leq A/h.$$

<u>Proof</u>

We retain the notations of section 2.2, Chapter II. Let K be a triangle of $\mathcal{T}_h$ and let $\hat{K}$ be its reference triangle, as in figure 3. By Lemma 2.1, Chapter II, we have

(2.36) $$|\mu|_{1,K} \leq C_1 \, \|B_K^{-1}\| \, |\det B_K|^{\frac{1}{2}} \, |\hat{\mu}|_{1,\hat{K}} \quad \forall \mu \in H^1(K).$$

As $\|B_K^{-1}\| \leq C_2/\rho_K$ and since $\mathcal{T}_h$ is uniformly regular, we find :

(2.37) $$\|B_K^{-1}\| \leq C_2 \sigma / h_K \leq C_2 \frac{\sigma}{\tau h}.$$

Substituting (2.37) in (2.36), we get :

(2.38) $$|\mu|_{1,K} \geq C_3 \frac{\sigma}{\tau h} \, |\det B_K|^{\frac{1}{2}} \, |\hat{\mu}|_{1,\hat{K}} \quad \forall \mu \in H^1(K).$$

Now, if $\hat{\mu} \in P_\ell$, which is a finite-dimensional space, there exists a constant C_4 that depends solely upon the geometry of $\hat{K}$ such that

$$|\hat{\mu}|_{1,\hat{K}} \leq C_4 \, \|\hat{\mu}\|_{0,\hat{K}}.$$

By applying again Lemma 2.1, Chapter II, we get :

$$|\hat{\mu}|_{1,\hat{K}} \leq C_5 |\det B_K|^{-\frac{1}{2}} \, \|\mu\|_{0,K}.$$

When substituted in (2.38), this yields :

$$|\mu|_{1,K} \leq C_6 \frac{\sigma}{\tau h} \|\mu\|_{0,K}.$$

Hence

$$\|\mu\|_{1,K} \leq \frac{1}{h}\left(\frac{C_6^2 \sigma^2}{\tau^2} + h^2\right)^{\frac{1}{2}} \|\mu\|_{0,K} \quad ,$$

thus proving (2.35). ■

<u>THEOREM 2.4.</u>

<u>Let $\mathcal{T}_h$ be a uniformly regular family of triangulations of $\bar{\Omega}$ and let the spaces</u> Φ_h, Θ_h <u>and</u> M_h <u>be defined by (2.33) and (2.34). Then, if</u> $\vec{u} = \overrightarrow{\text{curl}}\,\psi \in [H^{\ell+1}(\Omega)]^2$ <u>for an integer</u> $\ell \geq 2$, <u>we have the error bound :</u>

(2.39) $$\|\vec{u}-\vec{u}_h\|_{0,\Omega} + \|\omega-\omega_h\|_{0,\Omega} \leq C h^{\ell-1}\{|\vec{u}|_{\ell,\Omega} + |\vec{u}|_{\ell+1,\Omega}\}.$$

Proof

Since $\Phi_h \subset \Theta_h = M_h$, the bound (2.32) is valid, i.e. :

(2.40) $$\|\vec{u}-\vec{u}_h\|_{0,\Omega} + \|\omega-\omega_h\|_{0,\Omega} \leq C_1\Big[\big(1+S(h)\big)\Big\{\inf_{\phi_h\in\Phi_h} |\psi-\phi_h|_{1,\Omega} + \inf_{\theta_h\in\Theta_h} \|\omega-\theta_h\|_{0,\Omega}\Big\} + \inf_{\mu_h\in M_h} \|\nu\omega-\mu_h\|_{1,\Omega}\Big].$$

Therefore, we must derive an estimate for $|\psi-\phi_h|_{1,\Omega}$, $\|\omega-\theta_h\|_{0,\Omega}$ and $\|\nu\omega-\mu_h\|_{1,\Omega}$. Now, our spaces Θ_h, M_h and Φ_h are classical and there exists a standard interpolation operator $\Pi_h \in \mathcal{L}\big(H^2(\Omega);\Theta_h\big)$ such that

(2.41) $$|\theta-\Pi_h\theta|_{k,\Omega} \leq C_2 h^{m-k}|\theta|_{m,\Omega} \quad \forall\theta\in H^m(\Omega),$$

$$\text{for } 2\leq m\leq \ell+1 \text{ and } 0\leq k\leq m.$$

Then, if the stream function ψ belongs to $H^{\ell+1}(\Omega)$, we have the bound :

(2.42) $$\inf_{\phi_h\in\Phi_h} |\psi-\phi_h|_{1,\Omega} \leq C_2 h^{\ell}|\psi|_{\ell+1,\Omega}.$$

Next, if we assume that $\omega = -\Delta\psi \in H^{\ell}(\Omega)$, then

(2.43) $$\inf_{\theta_h\in\Theta_h} \|\omega-\theta_h\|_{0,\Omega} \leq C_2 h^{\ell}|\omega|_{\ell,\Omega}$$

and

(2.44) $$\inf_{\mu_h\in M_h} \|\nu\omega-\mu_h\|_{1,\Omega} \leq C_3 h^{\ell-1}|\omega|_{\ell,\Omega}.$$

Upon substituting (2.42), (2.43) and (2.44) in (2.40) and using (2.35), we get

$$\|\vec{u}-\vec{u}_h\|_{0,\Omega} + \|\omega-\omega_h\|_{0,\Omega} \leq C_4 h^{\ell-1}\{|\psi|_{\ell+1,\Omega} + |\omega|_{\ell,\Omega}\} \quad .$$ ∎

Remarks 2.3.

1°/ Note that the estimate (2.39) holds provided $\vec{u} = \overrightarrow{\text{curl}}\,\psi$ belongs to $[H^{\ell}(\Omega)]^2$ and $\omega = \text{curl}\,\vec{u}$ belongs to $H^{\ell}(\Omega)$.

2°/ When $\ell = 1$, we cannot derive the convergence of the above method from Theorem 2.4. A refined analysis proving the convergence and optimal error estimates can be found in [23]. Otherwise, it can be shown that the method is of order one when the triangulation is generated by three families of parallel lines (cf. [24]). ∎

CHAPTER IV

THE STATIONARY NAVIER-STOKES EQUATIONS

§ 1. A CLASS OF NON-LINEAR PROBLEMS

In this paragraph, we study a non-linear generalization of the abstract variational problem analyzed in §4, Chapter I. This family of non-linear problems contains, in particular, the Navier-Stokes problem.

We retain the notations of the linear problem. Namely, we consider two real Hilbert spaces X and M (normed respectively by $\|\cdot\|_X$ and $\|\cdot\|_M$) and a continuous bilinear form $b(v,\mu)$ on $X \times M$. The non-linearity is introduced by means of a form defined on $X \times X \times X$:

$$a(w;u,v)$$

where, for w in X, the mapping $(u,v) \mapsto a(w;u,v)$ is a bilinear and continuous form on $X \times X$. Then, for a given element ℓ of X', we consider the following problem:

$$(Q)\begin{cases} \text{Find a pair } (u,\lambda) \text{ in } X \times M \text{ satisfying} & \\ (1.1) \quad a(u;u,v) + b(v,\lambda) = \langle \ell, v \rangle & \forall v \in X, \\ (1.2) \quad b(u,\mu) = 0 & \forall \mu \in M. \end{cases}$$

We introduce the operators $A(w) \in \mathcal{L}(X;X')$, for w in X, and $B \in \mathcal{L}(X;M')$ defined by

$$\langle A(w)u,v\rangle = a(w;u,v) \quad \forall u,v \in X,$$

$$\langle Bv,\mu\rangle = b(v,\mu) \quad \forall v \in X, \quad \forall \mu \in M.$$

With these operators, problem (Q) becomes :

$$(Q)\begin{cases} \text{Find } u \in X \text{ and } \lambda \in M \text{ such that} \\ (1.3) \quad A(u)u + B'\lambda = \ell \text{ in } X' \\ (1.4) \quad Bu = 0 \text{ in } M'. \end{cases}$$

As in the linear case, we set $V = \operatorname{Ker} B$ in X ; then problem (P) associated with problem (Q) is :

$$
(P)\left\{\begin{array}{ll} & \text{Find } u \in V \text{ satisfying :} \\ (1.5) & a(u;u,v) = \langle \ell, v\rangle \quad \forall v \in V. \end{array}\right.
$$

f course, if (u,λ) is a solution of problem (Q), then u is a solution of pro-
lem (P). The converse can be established as in the linear case. And therefore,
e real difficulty here lies in solving the non-linear problem (P). To begin
th, we need a consequence of the classical fixed point Brouwer's theorem :

EOREM 1.1.

Let C denote a non-void, convex and compact subset of a finite-dimensional
ace and let F be a continuous mapping from C into C. Then F has at least one
xed point.

ROLLARY 1.1.

Let H be a *finite-dimensional* Hilbert space whose scalar product is denoted
$(\cdot,\cdot)$ and corresponding norm by $|\cdot|$. Let P be a continuous mapping from H
to H with the following property :
there exists $\xi > 0$ such that

$$
(P(f),f) > 0 \quad \forall f \in H \text{ with } |f| = \xi. \tag{.6}
$$

en, there exists an element f in H such that :

$$
|f| \leq \xi, \quad P(f) = 0. \tag{7}
$$

oof

The proof proceeds by contradiction. Suppose that $P(f) \neq 0$ in the sphere
$\{f \in H \; ; \; |f| \leq \xi\}$. Then the mapping

$$
f \mapsto -\xi \frac{P(f)}{|P(f)|}
$$

continuous from D into D. As the dimension of H is finite, and since D is
iously convex, compact and non-empty, we can apply Theorem 1.1 asserting that
re exists an f in D such that

$$
f = -\xi \frac{P(f)}{|P(f)|} .
$$

s, we have exhibited an f such that $|f| = \xi$ and

$$\big(P(f),f\big) = -\xi|P(f)| < 0,$$

since $f \in D$. This contradicts (1.6). ■

Now, we are in a position to establish the following existence result.

THEOREM 1.2

Assume that the following hypotheses hold :

(i) there exists a constant $\alpha > 0$ such that

$$(1.8) \qquad a(v;v,v) \geq \alpha \|v\|_X^2 \quad \forall v \in V ;$$

(ii) the space V is separable and the form $u \mapsto a(u;u,v)$ is weakly continuous in V, i.e.

$$(1.9) \quad u_m \rightarrow u \text{ weakly in } V \text{ (as } m \to \infty) \text{ implies that } \lim_{m\to\infty} a(u_m;u_m,v) = a(u;u,v) \ \forall v \in V .$$

Then, problem (P) has at least one solution u in V.

Proof

We shall construct a sequence of approximate solutions by Galerkin's method. Since V is separable, there exists a sequence $(w_i)_{i\geq 1}$ in V such that :

(i) for all $m \geq 1$, the elements $w_1, \dots, w_m$ are linearly independent,

(ii) the finite linear combinations of the w_i's, $\sum_i \xi_i w_i$, are dense in V.

Then, we denote by V_m the subspace of V spanned by $w_1, \dots, w_m$ and we approximate problem (P) by :

$$(P_m) \begin{cases} \text{Find } u_m \in V_m \text{ satisfying} \\ (1.10) \qquad a(u_m;u_m,v) = \langle \ell, v \rangle \quad \forall v \in V_m . \end{cases}$$

This is a system of m non-linear equations in m unknowns.

Our object now is to show that for each m, (P_m) has at least one solution u_m. Then, we shall construct a sequence (u_m) by taking for each m one of these solutions and establish that any such sequence (u_m) converges towards a solution of (P).

In order to prove the existence of u_m, we consider the operator $\mathcal{P}_m : V_m \mapsto V_m$, defined $\forall v \in V_m$, by :

$$(\mathcal{P}_m(v),w_i) = a(v;v,w_i) - <\ell,w_i> \quad \text{for} \quad 1 \leq i \leq m.$$

In particular,

$$(\mathcal{P}_m(v),v) = a(v;v,v) - <\ell,v> .$$

Thus, if we denote by $\|\ell\|^* = \sup_{v \in V} \frac{|<\ell,v>|}{\|v\|_X}$, the norm of ℓ in V', then hypothesis (1.8) implies that

$$(\mathcal{P}_m(v),v) \geq (\alpha \|v\|_X - \|\ell\|^*)\|v\|_X .$$

Hence, we choose $\xi > \frac{\|\ell\|^*}{\alpha}$ and $\forall v \in V_m$ such that $\|v\|_X = \xi$, we have :

$$(\mathcal{P}_m(v),v) > 0.$$

Moreover, $\mathcal{P}_m$ is continuous in V_m by virtue of hypothesis (1.9). As the dimension of V_m is finite, we can therefore make use of Corollary 1.1. Hence, there exists an element u_m of V_m that satisfies problem (P_m) and furthermore

$$0 = (\mathcal{P}(u_m),u_m) \geq (\alpha \|u_m\|_X - \|\ell\|^*)\|u_m\|_X ,$$

so that

$$\|u_m\|_X \leq \frac{\|\ell\|^*}{\alpha} . \tag{1.11}$$

Now, we examine the convergence of (u_m), as m tends to infinity. From (1.11), we see that the sequence (u_m) is bounded in V. Therefore, we can extract a subsequence (u_{m_p}) such that

$$u_{m_p} \to u \text{ weakly in } V \text{ as } p \to \infty.$$

Then, hypothesis (1.9) implies that

$$\lim_{p\to\infty} a(u_{m_p};u_{m_p},v) = a(u;u,v) \quad \forall v \in V .$$

Combined with (1.10), with large enough $m = m_p$, this yields :

$$a(u;u,w_i) = <\ell,w_i> \quad \forall i \geq 1.$$

Since each element of V is the limit of finite linear combinations of w_i, we get by density

$$a(u;u,v) = <\ell,v> \quad \forall v \in V .$$

Therefore, u is a solution of (P). ■

Now, we turn to the uniqueness of the solution. This requires stronger hypotheses than (1.8) and (1.9). Namely, we assume that

(i) form a is uniformly elliptic with respect to w ; that is, there exists a constant $\alpha>0$ such that :

$$a(w;v,v) \geqslant \alpha \|v\|_X^2 \quad \forall v,w \in V ; \tag{1.12}$$

(ii) the mapping $w \mapsto A(w)$ is locally Lipschitz continuous in V ; that is, there exists a continuous and monotonically increasing function $L : \mathbb{R}^+ \mapsto \mathbb{R}^+$ such that $\forall \xi > 0$:

$$|a(w_1;u,v) - a(w_2;u,v)| \leqslant L(\xi) \|u\|_X \|v\|_X \|w_1-w_2\|_X \quad \forall u,v \in V, \tag{1.13}$$

$\forall w_1, w_2 \in D_\xi$, where $D_\xi = \{v \in V ; \|v\|_X \leqslant \xi\}$.

With these hypotheses, we have the following uniqueness result.

THEOREM 1.3.

Suppose that (1.12) and (1.13) hold. Then, under the condition

$$\frac{\|\ell\|^\star}{\alpha^2} L\left(\frac{\|\ell\|^\star}{\alpha}\right) < 1, \tag{1.14}$$

problem (P) has a unique solution u in V.

Proof

According to hypothesis (1.12) and Lax-Milgram's theorem, it follows that, for each w in V, the operator $A(w) \in \mathcal{L}(V;V')$ is invertible. Moreover, its inverse operator, $T(w)$, belongs to $\mathcal{L}(V';V)$ and satisfies :

$$\|T(w)\|_{\mathcal{L}(V';V)} \leqslant \frac{1}{\alpha} . \tag{1.15}$$

With these notations, equation (1.5) of problem (P) becomes :

$$A(u)u = \ell \text{ in } V',$$

i.e.

$$u = T(u)\ell \text{ in } V.$$

Now, let us show that, owing to hypotheses (1.13) and (1.14), the mapping $v \mapsto T(v)\ell$ is a strict contraction from D_ξ into D_ξ with $\xi = \frac{\|\ell\|^\star}{\alpha}$. First, we check that $T(v)\ell$ belongs to D_ξ :

$$\|T(v)\ell\|_X \leq \|T(v)\|_{\mathcal{L}(V';V)}\|\ell\|^{\star} \leq \frac{1}{\alpha}\|\ell\|^{\star} = \xi.$$

Next, we evaluate $T(u)-T(v)$ for u and v in D_ξ. By virtue of the identity

$$T(u) - T(v) = T(u)\big(A(v) - A(u)\big)T(v)$$

and (1.15), we find :

$$\|T(u)-T(v)\|_{\mathcal{L}(V';V)} \leq \frac{1}{\alpha^2}\|A(v)-A(u)\|_{\mathcal{L}(V;V')} .$$

Therefore (1.13) yields :

$$\|T(u)\ell-T(v)\ell\|_X \leq \frac{1}{\alpha^2}\|\ell\|^{\star} L(\xi)\|v-u\|_X < \|v-u\|_X ,$$

thanks to (1.14). ■

Remark 1.1.

Since the mapping $v \mapsto T(v)\ell$ is a strict contraction in D_ξ, its fixed point can be computed by the method of successive approximations. More precisely, starting from any u_0 in D_ξ, we construct the sequence (u_m) by :

$$u_{m+1} = T(u_m)\ell,$$

or equivalently by :

$$a(u_m;u_{m+1},v) = <\ell,v> \quad \forall v\in V .$$

Then

$$\lim_{m\to\infty}\|u_m-u\|_X = 0.$$

Note also that u_0 need not be picked in D_ξ, since for any u_0 in V, $u_1\big(=T(u_0)\ell\big)$ is necessarily in D_ξ. ■

We end this paragraph by solving problem (Q). As mentioned at the beginning, we use the same argument as in the linear case.

THEOREM 1.4

Suppose that the form b satisfies the inf-sup condition :

$$(1.16) \qquad \sup_{v\in X}\frac{b(v,\mu)}{\|v\|_X} \geq \beta\|\mu\|_M \quad \forall\mu\in M, \text{ with } \beta>0.$$

Then, for each solution u of problem (P), there exists a unique λ in M such that the pair (u,λ) satisfies problem (Q).

Proof

According to (1.3) and (1.4) we must find u in V and λ in M such that

$$A(u)u + B'\lambda = \ell \text{ in } X'.$$

Now, if $u \in V$ is a solution of (P), then $\ell - A(u)u$ belongs to V^0, the polar set of V. And by virtue of (1.16), B' is an isomorphism from M onto V^0 (cf. Lemma 4.1, Chapter I). Thus there exists a unique λ such that (u,λ) is a solution of (1.3)■

Remark 1.2

Under the assumptions (1.12), (1.13), (1.14) and (1.16), we can extend the iterative scheme of remark 1.1 to solve problem (Q). More precisely, starting from any element u_0 in V, we construct the sequence $(u_m,\lambda_m) \in V \times M$, with $m \geq 1$, by solving the linear system :

$$(1.17) \qquad a(u_{m-1};u_m,v) + b(v,\lambda_m) = \langle \ell, v \rangle \quad \forall v \in X.$$

It can be proved that, whatever u_0 in V,

$$\lim_{m\to\infty} \{ \| u_m - u \|_X + \| \lambda_m - \lambda \|_M \} = 0. \qquad ■$$

§ 2. APPLICATION TO THE NAVIER-STOKES EQUATIONS

The general non linear theory of the preceding paragraph is applied here to the stationary Navier-Stokes equations. The major theoretical tools are recalled, without proof, in the first section.

2.1. SOME RESULTS OF FUNCTIONAL ANALYSIS

The notations used here are those of section 1.1, Chapter I. The first theorem is concerned with the Sobolev inequalities and the corresponding compactness properties of Sobolev spaces.

THEOREM 2.1.

Let $m \in \mathbb{N}$ with $m \geq 1$ and let $p \in \mathbb{R}$ with $1 \leq p \leq \infty$. If Ω is an open subset of $\mathbb{R}^n$

with a Lipschitz continuous boundary Γ, then the following imbeddings hold algebraically and topologically :

$$(2.1)\qquad W^{m,p}(\Omega) \subset \begin{cases} L^q(\Omega) & \text{provided } \frac{1}{q} = \frac{1}{p} - \frac{m}{n} > 0, \\ L^q_{loc}(\Omega) \ \ \forall q \text{ with } 1 \leqslant q < \infty & \text{provided } \frac{1}{p} = \frac{m}{n}, \\ \mathcal{C}^\circ(\bar{\Omega}) & \text{provided } \frac{1}{p} < \frac{m}{n}. \end{cases}$$

Moreover, if Ω is bounded, the canonical imbedding of $W^{1,p}(\Omega)$ into $L^{q_1}(\Omega)$ is compact $\forall q_1 \in \mathbb{R}$ that satisfy

$$(2.2)\qquad \begin{cases} 1 \leqslant q_1 < q & \text{whenever } \frac{1}{q} = \frac{1}{p} - \frac{1}{n} > 0, \\ \text{or} \quad 1 \leqslant q_1 < \infty & \text{when } p \geqslant n. \end{cases}$$

Remark 2.1.

The preceding theorem can be extended immediately to derivatives of functions. More precisely, if $v \in W^{m,p}(\Omega)$ and $\alpha \in \mathbb{N}^n$ with $|\alpha| \leqslant \ell \leqslant m$, then

$$\partial^\alpha v \in \begin{cases} L^q(\Omega) & \text{if } \frac{1}{q} = \frac{1}{p} - \frac{m-\ell}{n} > 0, \\ L^q_{loc}(\Omega) & \forall q \text{ if } \frac{1}{p} = \frac{m-\ell}{n}, \\ \mathcal{C}^\circ(\bar{\Omega}) & \text{if } \frac{1}{p} < \frac{m-\ell}{n}. \end{cases}$$

Similarly, if Ω is bounded, the canonical imbedding of $W^{m+1,p}(\Omega)$ into $W^{m,q_1}(\Omega)$ is compact when q_1 satisfies (2.2). ∎

In the case of Navier-Stokes equations, the dimension n is 2 or 3 and the space most often used is $H^1(\Omega)$. Theorem 2.1 gives the following algebraic and topological imbeddings :

$$\text{if } n = 2, \qquad H^1(\Omega) \subset L^q_{loc}(\Omega) \text{ for } 1 \leqslant q < \infty,$$

$$\text{if } n = 3, \qquad H^1(\Omega) \subset L^6(\Omega).$$

Furthermore, when Ω is bounded, the canonical imbedding of $H^1(\Omega)$ into $L^q(\Omega)$ is compact for :

$$1 \leqslant q < \infty \text{ when } n = 2$$

and

$$1 \leqslant q < 6 \text{ when } n = 3.$$

The conclusion of Theorem 2.1 is also valid for the fractional Sobolev space $H^s(\Omega)$ for $s \in \mathbb{R}^+$. Let us recall the following definitions :

$$H^s(\mathbb{R}^n) = \{v \in L^2(\mathbb{R}^n) \; ; \; \xi \mapsto (1+\xi^2)^{\frac{s}{2}} \hat{v}(\xi) \in L^2(\mathbb{R}^n_\xi)\},$$

normed by

$$\|v\|_{s,\mathbb{R}^n} = \left(\|v\|^2_{0,\mathbb{R}^n} + \|(1+\xi^2)^{\frac{s}{2}} \hat{v}(\xi)\|^2_{0,\mathbb{R}^n_\xi}\right)^{\frac{1}{2}} \quad ,$$

where $\hat{v}$ denotes the Fourier transform of v ; and

$$H^s(\Omega) = \{v \in L^2(\Omega) \; ; \; v \text{ is the restriction to } \Omega \text{ of a function } \bar{v} \in H^s(\mathbb{R}^n)\},$$

normed by

$$\|v\|_{s,\Omega} = \inf_{\bar{v} \in H^s(\mathbb{R}^n), \bar{v}|_\Omega = v} \|\bar{v}\|_{s,\mathbb{R}^n} \; ;$$

$H^s_0(\Omega)$ and $H^{-s}(\Omega)$ are defined exactly in the same way as in (1.4) and (1.5), Chapter I. When Γ is Lipschitz continuous and s is an integer, it can be shown that this definition yields the usual Sobolev space with, of course, an equivalent norm. As far as the Sobolev imbedding is concerned, the following result holds :

THEOREM 2.2.

Suppose that Γ is Lipschitz continuous. Then

$$H^s(\Omega) \subset L^p(\Omega)$$

algebraically and topologically, where

$$\frac{1}{p} = \frac{1}{2} - \frac{s}{n}, \text{ provided } \frac{1}{2} - \frac{s}{n} > 0. \tag{2.3}$$

Moreover, if Ω is bounded, the imbedding of $H^s(\Omega)$ into $L^{q_1}(\Omega)$ is compact $\forall q_1$ such that $1 \leq q_1 < p$ with p defined by (2.3).

Finally, the next theorem states briefly the main property of interpolation in Sobolev spaces.

THEOREM 2.3.

Let $\theta \in [0,1]$, let s_i and t_i be two pairs of real numbers with $0 \leq t_i \leq s_i$, for $i = 1,2$ and let $\mathcal{L}_i$ and $\mathcal{L}_\theta$ denote respectively $\mathcal{L}(H^{s_i}(\Omega); H^{t_i}(\Omega))$ and $\mathcal{L}(H^{(1-\theta)s_1+\theta s_2}(\Omega) \; ; \; H^{(1-\theta)t_1+\theta t_2}(\Omega))$. Let Π be an operator in $\mathcal{L}_1 \cap \mathcal{L}_2$; then Π also belongs to $\mathcal{L}_\theta$ and there exists a constant C :

$$\|\Pi\|_{\mathcal{L}_\theta} \leq C\|\Pi\|_{\mathcal{L}_1}^{1-\theta}\,\|\Pi\|_{\mathcal{L}_2}^{\theta}.$$

As an application of this last theorem, consider the solution u of the Dirichlet problem

$$-\Delta u = \ell \quad \text{in } \Omega,\ u|_\Gamma = 0.$$

We know that $u \in H_0^1(\Omega)$ when ℓ belongs to $H^{-1}(\Omega)$ and that $u \in H^2(\Omega) \cap H_0^1(\Omega)$ when ℓ is in $L^2(\Omega)$. Therefore, by interpolation we find that if $\ell \in H^{-s}(\Omega)$ for an $s \in [0,1]$, then

$$u \in H^{2-s}(\Omega) \cap H_0^1(\Omega).$$

2.2. SOLUTIONS OF THE NAVIER-STOKES PROBLEM

We have given in § 5.1, Chapter I the following "velocity-pressure" formulation of the stationary Navier-Stokes equations :

$$(2.4)\qquad \begin{cases} -\nu\Delta\vec{u} + \sum_{j=1}^{n} u_j \dfrac{\partial \vec{u}}{\partial x_j} + \overrightarrow{\text{grad}}\, p = \vec{f} \text{ in } \Omega, \\ \text{div}\,\vec{u} = 0 \text{ in } \Omega, \\ \vec{u}|_\Gamma = \vec{0}, \end{cases}$$

where again Ω is a bounded domain of $\mathbb{R}^n$ with a Lipschitz continuous boundary Γ. In order to write problem (2.4) in a variational form, we introduce the trilinear functional

$$(2.5)\qquad a_1(\vec{w};\vec{u},\vec{v}) = \sum_{i,j=1}^{n} \int_\Omega w_j \frac{\partial u_i}{\partial x_j} v_i \, dx,$$

and we get immediately :

$$\left(\sum_{j=1}^{n} u_j \frac{\partial \vec{u}}{\partial x_j}, \vec{v}\right) = a_1(\vec{u};\vec{u},\vec{v}).$$

We also recall the following spaces

$$\mathcal{V} = \{\vec{v} \in (\mathcal{D}(\Omega))^n \; ; \; \text{div}\,\vec{v} = 0\}$$

and

$$V = \{\vec{v} \in (H_0^1(\Omega))^n \; ; \; \text{div}\,\vec{v} = 0\}.$$

The next two lemmas state useful properties of a_1.

LEMMA 2.1

For $n \leq 4$, the trilinear form a_1 is continuous on $[(H^1(\Omega))^n]^3$.

Proof

According to Theorem 2.1, $H^1(\Omega)$ is continuously imbedded in $L^4(\Omega)$, when $n \leq 4$. Then, Hölder's inequality implies that

$$w_j \frac{\partial u_i}{\partial x_j} v_i \in L^1(\Omega) \quad \forall \vec{u},\vec{v},\vec{w} \in (H^1(\Omega))^n$$

and

$$\left| \int_\Omega w_j \frac{\partial u_i}{\partial x_j} v_i \, dx \right| \leq C_1 \| w_j \|_{1,\Omega} |u_i|_{1,\Omega} \| v_i \|_{1,\Omega}.$$

Thus the form a_1 is well defined and continuous on $[(H^1(\Omega))^n]^3$ and

$$(2.6) \qquad |a_1(\vec{w};\vec{u},\vec{v})| \leq C_2 \|\vec{w}\|_{1,\Omega} |\vec{u}|_{1,\Omega} \|\vec{v}\|_{1,\Omega}.$$

■

LEMMA 2.2

Let $\vec{u} \in (H^1(\Omega))^n$ with $\operatorname{div} \vec{u} = 0$ and $\gamma_\nu \vec{u} = 0$ and let $\vec{v}$ and $\vec{w} \in (H^1_0(\Omega))^n$; then we have :

$$(2.7) \qquad a_1(\vec{u};\vec{v},\vec{v}) = 0,$$

$$(2.8) \qquad a_1(\vec{u};\vec{v},\vec{w}) = -a_1(\vec{u};\vec{w},\vec{v}).$$

Proof

Clearly, it suffices to check (2.8). For this, let us take $\vec{u}$ in $\mathcal{V}$ and $\vec{v}$ and $\vec{w}$ in $(\mathcal{D}(\Omega))^n$; we have :

$$\int_\Omega u_j \left(w_i \frac{\partial v_i}{\partial x_j} + v_i \frac{\partial w_i}{\partial x_j} \right) dx = \int_\Omega u_j \frac{\partial}{\partial x_j} (v_i w_i) dx$$

$$= - \int_\Omega \frac{\partial u_j}{\partial x_j} v_i w_i \, dx.$$

Hence

$$a_1(\vec{u};\vec{v},\vec{w}) + a_1(\vec{u};\vec{w},\vec{v}) = 0.$$

Then we derive (2.8) by density. ■

Now, let

$$a_0(\vec{u},\vec{v}) = \nu(\overrightarrow{\text{grad}}\, \vec{u}, \overrightarrow{\text{grad}}\, \vec{v}),$$

$$(2.9) \qquad a(\vec{w};\vec{u},\vec{v}) = a_0(\vec{u},\vec{v}) + a_1(\vec{w};\vec{u},\vec{v}).$$

Then problem (2.4) has the equivalent variational form :

$$(2.10) \quad \begin{cases} \text{Find a pair } (\vec{u},p) \text{ in } V \times L_0^2(\Omega) \text{ such that} \\ a(\vec{u};\vec{u},\vec{v}) - (p, \operatorname{div} \vec{v}) = (\vec{f},\vec{v}) \quad \forall \vec{v} \in \left(H_0^1(\Omega)\right)^n. \end{cases}$$

THEOREM 2.4

Let $n \leqslant 4$ and let Ω be a bounded domain of $\mathbb{R}^n$ with a Lipschitz continuous boundary Γ. Given a function $\vec{f}$ in $\left(H^{-1}(\Omega)\right)^n$, there exists at least one pair $(\vec{u},p)$ in $V \times L_0^2(\Omega)$ that satisfies (2.10).

Proof

We apply the material of §1 as follows. With $X = \left(H_0^1(\Omega)\right)^n$ normed by $|\cdot|_{1,\Omega}$, $M = L_0^2(\Omega)$, $a(\vec{w};\vec{u},\vec{v})$ given by (2.9),

$$b(\vec{v},q) = -(q, \operatorname{div} \vec{v})$$

and

$$\langle \ell,\vec{v} \rangle = \langle \vec{f},\vec{v} \rangle ,$$

we can consider that (2.10) is a particular case of problem (Q) of §1. Thus, we must check the hypotheses of Theorems 1.2 and 1.4. First, by virtue of (2.7), we get

$$a(\vec{u};\vec{v},\vec{v}) = a_0(\vec{v},\vec{v}) = \nu|\vec{v}|_{1,\Omega}^2 .$$

Therefore, a satisfies property (1.12).

Next, let $\vec{u}$ be a function of V and $\vec{u}_m$ a sequence in V such that $u_m \to u$ weakly in V as $m \to \infty$. Then Theorem 2.1 implies that

$$\lim_{m\to\infty} \vec{u}_m = \vec{u} \text{ in } \left(L^2(\Omega)\right)^n.$$

Now, let $\vec{v} \in \mathcal{V}$ and let us take the limit of $a(\vec{u}_m;\vec{u}_m,\vec{v})$. According to (2.8), we have

$$a_1(\vec{u}_m;\vec{u}_m,\vec{v}) = -a_1(\vec{u}_m;\vec{v},\vec{u}_m) = -\left(\sum_{j=1}^{n} u_{mj} \frac{\partial \vec{v}}{\partial x_j}, \vec{u}_m\right).$$

As $\dfrac{\partial \vec{v}}{\partial x_j} \in \left(L^\infty(\Omega)\right)^n$ and $\lim\limits_{m\to\infty} u_{mj}\, u_{mi} = u_j\, u_i$ in $L^1(\Omega)$, it follows that

$$\lim_{m\to\infty} a_1(\vec{u}_m;\vec{u}_m,\vec{v}) = -\left(\sum_{j=1}^{n} u_j \frac{\partial \vec{v}}{\partial x_j},\vec{u}\right) = a_1(\vec{u};\vec{u},\vec{v}).$$

Besides that, it is clear that $\lim_{m\to\infty} a_0(\vec{u}_m,\vec{v}) = a_0(\vec{u},\vec{v})$. Therefore,

$$\lim_{m\to\infty} a(\vec{u}_m;\vec{u}_m,\vec{v}) = a(\vec{u};\vec{u},\vec{v}) \qquad \forall \vec{v}\in V$$

by virtue of the density of $\mathcal{V}$ in V and the trilinearity of a.

Thus the hypotheses of Theorem 1.2 are satisfied and therefore problem (2.10) has at least one solution $\vec{u}$ in V.

As far as the pressure is concerned, we have already seen in §5, Chapter I that the form b satisfies the inf-sup condition of Theorem 1.4. Therefore, for each solution $\vec{u}$ of (2.10) there exists a unique p in $L_0^2(\Omega)$ such that $(\vec{u},p)$ satisfies (2.10). ■

In the sequel, we shall concentrate on the approximation of nonsingular solutions of the Navier-Stokes equations. This concept defined below is a convenient sufficient condition for local uniqueness.

DEFINITION 2.1

Let X and Y be two Banach spaces, F a differentiable mapping from X into Y, F' its derivative, and let $u\in X$ be a solution of the equation $F(u)=0$. We say that u is a nonsingular solution if there exists a constant $\gamma>0$ such that

$$\|F'(u)\cdot v\|^{*}_{Y'} \geq \gamma\|v\|_X \qquad \forall v\in X .$$

In the Navier-Stokes case, the mapping $F: V \mapsto V'$ is defined by :

$$<F(\vec{u}),\vec{v}> = a(\vec{u};\vec{u},\vec{v}) - <\vec{f},\vec{v}> \qquad \forall \vec{u},\vec{v}\in V .$$

Clearly, F is everywhere differentiable in V and its derivative $F'(\vec{u})\in\mathcal{L}(V;V')$ is given by :

$$<F'(\vec{u})\vec{v},\vec{w}> = a_0(\vec{v},\vec{w}) + a_1(\vec{u};\vec{v},\vec{w}) + a_1(\vec{v};\vec{u},\vec{w}).$$

Thus, if $c(\vec{u};\vec{v},\vec{w})$ denotes the trilinear form defined on X^3 by :

$$c(\vec{u};\vec{v},\vec{w}) = a_0(\vec{v},\vec{w}) + a_1(\vec{u};\vec{v},\vec{w}) + a_1(\vec{v};\vec{u},\vec{w}), \tag{2.11}$$

then $\vec{u}_0 \in V$ is a nonsingular solution of the Navier-Stokes equations if and only if there exists a constant $\gamma > 0$ such that

$$\text{(2.12)} \qquad \sup_{\vec{v} \in V} \frac{|c(\vec{u}_0;\vec{u},\vec{v})|}{|\vec{v}|_{1,\Omega}} \geq \gamma |\vec{u}|_{1,\Omega} \quad \forall \vec{u} \in V .$$

This definition amounts to say that the problem :

$$\text{(2.13)} \qquad \begin{cases} \text{Find } \vec{u} \in V \text{ such that} \\ \qquad c(\vec{u}_0;\vec{u},\vec{v}) = \langle \vec{g},\vec{v} \rangle \quad \forall \vec{v} \in V \end{cases}$$

is well posed.

In order to study the nonsingular solutions of the Navier-Stokes equations, it will be very useful to introduce the operator $K = K_{\vec{u}_0} \in \mathcal{L}(X;V)$, depending upon a parameter $\vec{u}_0 \in X$, and defined as follows :

for each $\vec{u} \in X$, $K\vec{u}$ is the solution in V of the Stokes problem

$$\text{(2.14)} \qquad a_0(K\vec{u},\vec{v}) = a_1(\vec{u}_0;\vec{u},\vec{v}) + a_1(\vec{u};\vec{u}_0,\vec{v}) \quad \forall \vec{v} \in V .$$

It follows from (2.11) and (2.14) that :

$$\text{(2.15)} \qquad c(\vec{u}_0;\vec{u},\vec{v}) = a_0\big((I + K_{\vec{u}_0})\vec{u},\vec{v}\big) \quad \forall \vec{u}_0,\vec{u} \in X, \quad \forall \vec{v} \in V .$$

The importance of K appears in the next two lemmas.

<u>LEMMA 2.3</u>

<u>Suppose that</u> Γ <u>is of class</u> $\mathcal{C}^2$ <u>(or, if</u> Γ <u>is only Lipschitz continuous, suppose that</u> $n = 2$ <u>and</u> Ω <u>is convex). Then, for</u> $n \leq 4$, <u>the operator</u> K <u>is compact from</u> X <u>into</u> V.

<u>Proof</u>

Let $\vec{u}$ and $\vec{u}_0 \in X$. According to (2.14), $K\vec{u}$ is the solution of

$$\vec{z} \in V, \ a_0(\vec{z},\vec{v}) = (\vec{g},\vec{v}) \quad \forall \vec{v} \in V ,$$

where

$$\vec{g} = \sum_{j=1}^{n} \{u_j^0 \frac{\partial \vec{u}}{\partial x_j} + u_j \frac{\partial \vec{u}_0}{\partial x_j}\}.$$

In this expression, $\frac{\partial u_i}{\partial x_j} \in L^2(\Omega)$ and $u_j^0 \in H_0^1(\Omega) \subset L^4(\Omega)$ when $n \leq 4$, by virtue of (2.1). Hence $\vec{g} \in \big(L^{\frac{4}{3}}(\Omega)\big)^n$. As the hypotheses of Theorem 5.2, Chapter I, are satisfied, we infer that $K\vec{u} \in (W^{2,4/3}(\Omega))^n$ and

$$\|K\vec{u}\|_{2,4/3,\Omega} \leq C|\vec{u}_0|_{1,\Omega}|\vec{u}|_{1,\Omega}.$$

Then the compactness of K follows from the fact that the canonical imbedding of $W^{2,4/3}(\Omega)$ into $H^1(\Omega)$ is compact (cf. remark 2.1). ∎

LEMMA 2.4.

Let the hypotheses of Lemma 2.3 be satisfied. If u is a nonsingular solution of (2.10), then the operator $I + K_{\vec{u}_0}$ is invertible in $\mathcal{L}(X;X)$ or in $\mathcal{L}(V;V)$ and has a continuous inverse.

Proof

As $K_{\vec{u}_0}$ is compact, we can apply Fredholm's alternative to $I + K_{\vec{u}_0}$ (cf. Yosida [46]). Then equation (2.15) and the fact that problem (2.13) has a unique solution necessarily imply that $I + K_{\vec{u}_0}$ is invertible and has a continuous inverse. ∎

Now, we turn to the global uniqueness of the solution. For this, we introduce the norm of a_1 in V^3 :

$$N = \sup_{\vec{u},\vec{v},\vec{w} \in V} \frac{|a_1(\vec{w};\vec{u},\vec{v})|}{|\vec{u}|_{1,\Omega}|\vec{v}|_{1,\Omega}|\vec{w}|_{1,\Omega}}. \tag{2.16}$$

THEOREM 2.5.

Under the hypotheses of Theorem 2.4 and if in addition

$$\frac{N}{\nu^2}\|\ell\|^{\star} < 1, \tag{2.17}$$

then the Navier-Stokes problem (2.10) has a unique solution $(\vec{u},p)$ in $V \times L_0^2(\Omega)$.

Proof

Here again, we make use of §1, and more precisely of Theorem 1.3. We have already proved property (1.12) with $\alpha = \nu$ and it suffices to establish (1.13). Let $\vec{u},\vec{v},\vec{w}_1$ and $\vec{w}_2 \in V$; we have :

$$|a(\vec{w}_1;\vec{u},\vec{v}) - a(\vec{w}_2;\vec{u},\vec{v})| = |a_1(\vec{w}_1-\vec{w}_2;\vec{u},\vec{v})|$$
$$\leq N|\vec{w}_1-\vec{w}_2|_{1,\Omega}|\vec{u}|_{1,\Omega}|\vec{v}|_{1,\Omega}.$$

Therefore, a satisfies (1.13) with $L(\xi) = N \quad \forall\xi$. Then condition (2.17) coincides precisely with (1.14). Hence the conclusion of Theorem 1.3 is valid. ■

Remark 2.2.

In view of Remark 1.2, we see that the conditions of Theorem 2.5 are sufficient to guarantee that the iterative scheme :

$$(2.18)\quad \begin{cases} (\vec{u}^m, p^m) \in V \times L_0^2(\Omega) \quad \text{satisfying} \\ a(\vec{u}^{m-1}; \vec{u}^m, \vec{v}) - (p^m, \operatorname{div} \vec{v}) = \langle \vec{f}, \vec{v} \rangle \quad \forall \vec{v} \in \left(H_0^1(\Omega)\right)^n \end{cases},$$

starting from an arbitrary $\vec{u}^0$ in V, defines uniquely a sequence $(\vec{u}^m, p^m)$ in $V \times L_0^2(\Omega)$ such that

$$\lim_{m\to\infty} \left(|\vec{u}^m - \vec{u}|_{1,\Omega} + \|p^m - p\|_{0,\Omega}\right) = 0. \quad ■$$

Remark 2.3.

Condition (2.17) implies that the solution $\vec{u}$ is not only unique but also nonsingular. Indeed, by virtue of (2.7) and (2.16), we have :

$$c(\vec{u}; \vec{v}, \vec{v}) = \nu |\vec{v}|^2_{1,\Omega} + a_1(\vec{v}; \vec{u}, \vec{v})$$

$$\geq \left(\nu - N|\vec{u}|_{1,\Omega}\right) |\vec{v}|^2_{1,\Omega} \quad \forall \vec{v} \in V.$$

But since $|\vec{u}|_{1,\Omega} \leq \frac{1}{\nu} \|\vec{f}\|^*$, this implies that

$$c(\vec{u}; \vec{v}, \vec{v}) \geq \left(\nu - \frac{N}{\nu} \|\vec{f}\|^*\right) |\vec{v}|^2_{1,\Omega} \quad \forall \vec{v} \in V \; .$$

Thus, owing to (2.17), the bilinear form $(\vec{v}, \vec{w}) \mapsto c(\vec{u}; \vec{v}, \vec{w})$ is V-elliptic and a fortiori satisfies (2.12). ■

Let us now examine the two-dimensional case. Recall the space of stream functions associated with V :

$$\Phi = \{\phi \in H^2(\Omega) \; ; \; \phi|_{\Gamma_0} = 0, \; \phi|_{\Gamma_i} = \text{an arbitrary constant } c_i,$$

$$1 \leq i \leq p, \; \frac{\partial \phi}{\partial \nu}\Big|_{\Gamma} = 0\}.$$

In terms of stream functions, problem (2.10) reads as follows :

$$(2.19)\quad \begin{cases} \text{Find } \psi\in\Phi \text{ satisfying :} \\ \nu(\Delta\psi,\Delta\phi) + \int_\Omega \Delta\psi\left(\frac{\partial\psi}{\partial x_2}\frac{\partial\phi}{\partial x_1} - \frac{\partial\psi}{\partial x_1}\frac{\partial\phi}{\partial x_2}\right)dx = \langle\vec{f},\overrightarrow{\text{curl}}\,\phi\rangle \quad \forall\phi\in\Phi \ . \end{cases}$$

The next theorem establishes that these two problems are indeed equivalent.

THEOREM 2.6

Problems (2.10) and (2.19) are equivalent in the sense that if $(\vec{u},p)$ is a solution of (2.10) then *the* stream function $\psi\in\Phi$ of $\vec{u}$ satisfies (2.19) ; conversely if ψ is a solution of (2.19), then there exists exactly one element p of $L_0^2(\Omega)$ such that the pair $(\vec{u}=\overrightarrow{\text{curl}}\,\psi,p)$ satisfies (2.10).

Proof

Recall that $V=\{\overrightarrow{\text{curl}}\,\phi;\ \phi\in\Phi\}$ and that

$$(\overrightarrow{\text{grad}}\,\vec{u},\overrightarrow{\text{grad}}\,\vec{v}) = (\Delta\psi,\Delta\phi), \quad \forall\vec{u},\vec{v}\in V,$$

where $\overrightarrow{\text{curl}}\,\psi=\vec{u}$ and $\overrightarrow{\text{curl}}\,\phi=\vec{v}$. This takes care of the viscous term.

Now, we look at the convection term ; we have :

$$(2.20)\qquad a_1(\vec{u};\vec{u},\vec{v}) = \int_\Omega \text{curl}\,\vec{u}\,(u_1v_2-u_2v_1)dx \quad \forall\vec{u},\vec{v}\in V \ .$$

Indeed, for $\vec{u}$ and $\vec{v}$ in $\mathcal{V}$, we get immediately :

$$u_1\frac{\partial u_1}{\partial x_1}+u_2\frac{\partial u_1}{\partial x_2}=\frac{1}{2}\frac{\partial}{\partial x_1}(u_1^2+u_2^2)-u_2\,\text{curl}\,\vec{u},$$

$$u_1\frac{\partial u_2}{\partial x_1}+u_2\frac{\partial u_2}{\partial x_2}=\frac{1}{2}\frac{\partial}{\partial x_2}(u_1^2+u_2^2)+u_1\,\text{curl}\,\vec{u}.$$

Thus, by integration by parts, this yields :

$$a_1(\vec{u};\vec{u},\vec{v}) = -\frac{1}{2}(u_1^2+u_2^2,\text{div}\,\vec{v}) - \int_\Omega \text{curl}\,\vec{u}\,(u_2v_1-u_1v_2)dx.$$

Hence (2.20) holds for all $\vec{u}$ and $\vec{v}$ in $\mathcal{V}$, and by density for all $\vec{u}$ and $\vec{v}$ in V.

Therefore, the convection term can also be expressed as :

$$(2.21)\qquad a_1(\vec{u};\vec{u},\vec{v}) = \int_\Omega \Delta\psi\left(\frac{\partial\psi}{\partial x_2}\frac{\partial\phi}{\partial x_1} - \frac{\partial\psi}{\partial x_1}\frac{\partial\phi}{\partial x_2}\right)dx.$$

It follows that the problem :

$$(2.22)\quad \begin{cases} \text{Find } \vec{u} \text{ in } V \text{ satisfying} \\ \qquad a(\vec{u};\vec{u},\vec{v}) = \langle\vec{f},\vec{v}\rangle \quad \forall\vec{v}\in V \end{cases}$$

is equivalent to (2.19). ■

It remains to interpret problem (2.19). It can be easily checked by integration by parts that ψ satisfies the following equations :

$$(2.23)\quad \begin{cases} \nu\Delta^2\psi - \dfrac{\partial}{\partial x_1}\left(\Delta\psi\dfrac{\partial\psi}{\partial x_2}\right) + \dfrac{\partial}{\partial x_2}\left(\Delta\psi\dfrac{\partial\psi}{\partial x_1}\right) = \operatorname{curl}\vec{f}, \\ \psi|_{\Gamma_0} = 0,\ \psi|_{\Gamma_i} = \text{a constant } c_i,\ 1 \leq i \leq p \\ \dfrac{\partial\psi}{\partial\nu}\Big|_{\Gamma} = 0,\ \displaystyle\int_{\Gamma_i}\left(\nu\frac{\partial}{\partial n}\Delta\psi - \gamma_\tau\vec{f}\right)d\sigma = 0,\quad 1 \leq i \leq p, \end{cases}$$

where the last equation is formal and $\frac{\partial}{\partial n}$ denotes the normal derivative.

§3. A FIRST METHOD FOR APPROXIMATING THE NAVIER-STOKES EQUATIONS

The remaining two paragraphs are devoted to finite element approximations of the Navier-Stokes problem in two dimensions. In fact, the results that we establish can be extended to the discretization of the general non linear problems described in §1, but for the sake of simplicity we prefer to restrict the discussion to Navier-Stokes equations. In particular, we retain the notations of §2.

The finite element method proposed here is the one we introduced in Chapter II, §2 for solving the Stokes problem. But now, the situation is more complicated and we split the discussion into two cases according that the solution is unique or not.

3.1. THE UNIQUENESS CASE

In this section, the solution $\vec{u}$ is supposed to be unique and, more precisely, we assume that

$$(3.1)\qquad \frac{N}{\nu^2}\|\vec{f}\|^{\star} \leq 1-\delta \text{ for some } \delta > 0.$$

For each h, let W_h and M_h be two finite-dimensional spaces such that :

$$W_h \subset H^1(\Omega),\quad M_h \subset L_0^2(\Omega) = M.$$

Next, we set

$$W_{h,0} = W_h \cap H_0^1(\Omega),\quad X_h = (W_{h,0})^n \subset X.$$

Again, the space V_h corresponding to the form b is defined by :

$$V_h = \{\vec{v}_h \in X_h \; ; \; (q_h, \operatorname{div} \vec{v}_h) = 0 \quad \forall q_h \in M_h\}.$$

As usual, $V_h \not\subset V$ and in particular, the functions of V_h are not divergence-free. Hence some care must be taken in order to preserve the antisymmetry of form a_1, which plays a fundamental part in this section (cf. Lemma 2.2). For this purpose, the simplest thing is to introduce a slight variant of a_1 :

$$\tilde{a}_1(\vec{u};\vec{v},\vec{w}) = \frac{1}{2}\{a_1(\vec{u};\vec{v},\vec{w}) - a_1(\vec{u};\vec{w};\vec{v})\}.$$

By virtue of (2.8), it is clear that $\tilde{a}_1$ and a_1 coincide on $V \times X \times X$ and, of course, $\tilde{a}_1$ is antisymmetric with respect to its last two arguments. Therefore, the results of §2 are still valid if we replace everywhere a_1 by $\tilde{a}_1$. Here, we work exclusively with $\tilde{a}_1$ but in order to avoid a multiplicity of notations, we agree to drop the tilde.

Finally, we define the following discrete norms :

$$(3.2) \qquad N_h = \sup_{\vec{u}_h, \vec{v}_h, \vec{w}_h \in V_h} \frac{|a_1(\vec{w}_h;\vec{u}_h,\vec{v}_h)|}{|\vec{u}_h|_{1,\Omega} |\vec{v}_h|_{1,\Omega} |\vec{w}_h|_{1,\Omega}}$$

and

$$\|\vec{f}\|_h^* = \sup_{\vec{v}_h \in V_h} \frac{|\langle \vec{f}, \vec{v}_h \rangle|}{|\vec{v}_h|_{1,\Omega}}.$$

With the above notation, the discrete analogue of problem (2.10) is :

$$(Q_h) \begin{cases} \text{Find a pair } (\vec{u}_h, p_h) \text{ in } V_h \times M_h \text{ such that} \\ (3.3) \qquad a(\vec{u}_h;\vec{u}_h,\vec{v}_h) - (p_h, \operatorname{div} \vec{v}_h) = \langle \vec{f}, \vec{v}_h \rangle \quad \forall \vec{v}_h \in X_h. \end{cases}$$

Its corresponding problem (P_h) is :

$$(P_h) \begin{cases} \text{Find } \vec{u}_h \text{ in } V_h \text{ satisfying} \\ (3.4) \qquad a(\vec{u}_h;\vec{u}_h,\vec{v}_h) = \langle \vec{f}, \vec{v}_h \rangle \quad \forall \vec{v}_h \in V_h. \end{cases}$$

Just like in the linear case, we begin by examining problem (P_h) and showing the convergence of $\vec{u}_h$ toward $\vec{u}$.

<u>THEOREM 3.1</u>

<u>Problem</u> (P_h) <u>has at least one solution</u> $\vec{u}_h$ <u>in</u> V_h. <u>Moreover , if</u>

$$(3.5)\qquad \frac{N_h}{\nu^2}\|\vec{f}\|_h^{\star} < 1,$$

then this solution is unique and is the limit of the sequence defined by the iterative scheme :

$$(3.6)\qquad a(\vec{u}_h^m;\vec{u}_h^{m+1},\vec{v}_h) = \langle\vec{f},\vec{v}_h\rangle \quad \forall\vec{v}_h \in V_h$$

starting with an arbitrary $\vec{u}_h^0$ in V_h.

We skip the proof since it is a simplified version of the proof established in the continuous case.

Before we turn to the convergence of $\vec{u}_h$, let us recall briefly the approximation properties of X_h and M_h that we assumed in Chapter II, § 2.

Hypothesis H1

There exists a mapping $r_h \in \mathcal{L}\left((H^2(\Omega)\cap H_0^1(\Omega))^n;X_h\right)$ and an integer ℓ such that :

$$(3.7)\qquad \left(q_h,\operatorname{div}(\vec{v}-r_h\vec{v})\right) = 0 \quad \forall q_h \in M_h,$$

$$(3.8)\qquad \|r_h\vec{v}-\vec{v}\|_{1,\Omega} \leqslant C\,h^m\|\vec{v}\|_{m+1,\Omega} \quad \forall\vec{v}\in\left(H^{m+1}(\Omega)\cap H_0^1(\Omega)\right)^n,$$

$$\forall\, m\in\mathbb{N} \quad \text{with} \quad 1\leqslant m\leqslant \ell \;. \blacksquare$$

Hypothesis H2

The orthogonal projection operator ρ_h on M_h satisfies :

$$(3.9)\qquad \|q-\rho_h q\|_{0,\Omega} \leqslant C\,h^m\|q\|_{m,\Omega} \quad \forall q\in H^m(\Omega)\cap L_0^2(\Omega),\ 0\leqslant m\leqslant\ell \;. \blacksquare$$

With these hypotheses, we can prove the following result.

LEMMA 3.1.

If Hypotheses H1 and H2 are satisfied, then

$$\lim_{h\to o} N_h = N, \quad \lim_{h\to o}\|\vec{f}\|_h^{\star} = \|\vec{f}\|^{\star}.$$

Proof.

We just prove the first assertion, since the proof of the second statement is the same. As the dimension of V_h is finite, there exist functions $\vec{u}_h$, $\vec{v}_h$ and $\vec{w}_h$ in V_h such that

(3.10) $$|\vec{u}_h|_{1,\Omega} = |\vec{v}_h|_{1,\Omega} = |\vec{w}_h|_{1,\Omega} = 1$$

and

(3.11) $$|a_1(\vec{w}_h;\vec{u}_h,\vec{v}_h)| = N_h.$$

Also, by virtue of (2.6), N_h is bounded as h tends to zero. Therefore, we can extract a subsequence $(h_\rho)_{\rho\geq 1}$ such that

(3.12) $$\vec{u}_{h_\rho} \to \vec{u}^\star,\ \vec{v}_{h_\rho} \to \vec{v}^\star,\ \vec{w}_{h_\rho} \to \vec{w}^\star \text{ weakly in } \left(H_0^1(\Omega)\right)^n \text{ as } \rho\to\infty$$

and

$$\lim_{\rho\to\infty} N_{h_\rho} = N^\star.$$

Let us show that $N^\star = N$. The proof proceeds in two steps.

1°/ First, we prove that $N^\star \leq N$.

We remark that $\vec{u}^\star$, $\vec{v}^\star$ and $\vec{w}^\star$ are elements of V. Indeed, by virtue of (3.12), (3.9) and the definition of V_h, we have

$$(\operatorname{div}\vec{u}^\star, q) = 0 \quad \forall q \in H^1(\Omega),$$

and hence, by density $\operatorname{div}\vec{u}^\star = 0$. Therefore $\vec{u}^\star \in V$ and the same holds for $\vec{v}^\star$ and $\vec{w}^\star$.

Now, since the dimension $n \leq 3$, $H_0^1(\Omega)$ is compactly imbedded into $L^4(\Omega)$. Therefore (3.12) implies that

$$\lim_{\rho\to\infty} \vec{u}_{h_\rho} = \vec{u}^\star,\ \lim_{\rho\to\infty} \vec{v}_{h_\rho} = \vec{v}^\star,\ \lim_{\rho\to\infty} \vec{w}_{h_\rho} = \vec{w}^\star \text{ in } \left(L^4(\Omega)\right)^n.$$

Hence

$$\lim_{\rho\to\infty} \int_\Omega (w_{h_\rho})_j \frac{\partial (u_{h_\rho})_i}{\partial x_j} (v_{h_\rho})_i\, dx = \int_\Omega w_j^\star \frac{\partial u_i^\star}{\partial x_j} v_i^\star\, dx.$$

As this is a general term in the expression of a_1, it follows that

(3.13) $$\lim_{\rho\to\infty} |a_1(\vec{w}_{h_\rho};\vec{u}_{h_\rho},\vec{v}_{h_\rho})| = |a_1(\vec{w}^\star;\vec{u}^\star,\vec{v}^\star)| = N^\star.$$

Finally, from (3.10) and the lower semi-continuity of the norm for the weak topology, we derive the upper bounds

$$|\vec{u}^\star|_{1,\Omega} \leq 1,\ |\vec{v}^\star|_{1,\Omega} \leq 1,\ |\vec{w}^\star|_{1,\Omega} \leq 1.$$

Together with (3.13), this yields

$$N^\star \leq \frac{|a_1(\vec{w}^\star;\vec{u}^\star,\vec{v}^\star)|}{|\vec{w}^\star|_{1,\Omega}|\vec{v}^\star|_{1,\Omega}|\vec{u}^\star|_{1,\Omega}} \leq N,$$

thus proving the first inequality.

2°/ Next, we prove that $N^\star \geq N$.

Let $\vec{u}$, $\vec{v}$ and $\vec{w}$ belong to $\mathcal{V}$. From (3.8), we infer that :

$$\lim_{h\to o} r_h\vec{u} = \vec{u},\ \lim_{h\to o} r_h\vec{v} = \vec{v},\ \lim_{h\to o} r_h\vec{w} = \vec{w} \text{ in } V.$$

Whence

$$\lim_{h\to o} \frac{a_1(r_h\vec{w};r_h\vec{u},r_h\vec{v})}{|r_h\vec{u}|_{1,\Omega}|r_h\vec{v}|_{1,\Omega}|r_h\vec{w}|_{1,\Omega}} = \frac{a_1(\vec{w};\vec{u},\vec{v})}{|\vec{u}|_{1,\Omega}|\vec{v}|_{1,\Omega}|\vec{w}|_{1,\Omega}}.$$

Thus

$$N^\star \geq \frac{|a_1(\vec{w};\vec{u},\vec{v})|}{|\vec{u}|_{1,\Omega}|\vec{v}|_{1,\Omega}|\vec{w}|_{1,\Omega}} \quad \forall \vec{u},\vec{v},\vec{w} \in \mathcal{V}$$

and owing to the density of $\mathcal{V}$ in V, this means that $N^\star \geq N$.

Therefore $\lim_{\rho\to\infty} N_{h_\rho} = N$. The uniqueness of this limit implies the convergence of the entire sequence, i.e.

$$\lim_{h\to o} N_h = N. \qquad \blacksquare$$

THEOREM 3.2

Under Hypotheses H1 and H2, Condition (3.1) and if h is sufficiently small then problem (P_h) has a unique solution $\vec{u}_h$ in V_h and

$$\lim_{h\to o} |\vec{u}-\vec{u}_h|_{1,\Omega} = 0. \tag{3.14}$$

In addition, if the solution $(\vec{u},p)$ belongs to $(H^{m+1}(\Omega))^n \times (H^m(\Omega)\cap L_0^2(\Omega))$ for $m \leq \ell$, we have the following estimate :

$$|\vec{u}-\vec{u}_h|_{1,\Omega} \leq C\,h^m\,(\|\vec{u}\|_{m+1,\Omega} + \|p\|_{m,\Omega}). \tag{3.15}$$

Proof

According to Lemma 3.1 and Condition (3.1), we can choose h sufficiently small so that

$$\frac{N_h}{\nu^2}\|\vec{f}\|_h^\star \leq 1 - \frac{\delta}{2}. \tag{3.16}$$

Then, by virtue of Theorem 3.1, problem (P_h) has exactly one solution $\vec{u}_h$ in V_h. Let $\vec{v}_h$ be an arbitrary element of V_h, let $\vec{w}_h = \vec{u}_h - \vec{v}_h$ and consider the expression:

$$E_h = a(\vec{u}_h;\vec{u}_h,\vec{w}_h) - a(\vec{v}_h;\vec{v}_h,\vec{w}_h).$$

Let us derive a lower bound for E_h. Expanding, we get :

$$E_h = a_0(\vec{w}_h,\vec{w}_h) + a_1(\vec{u}_h;\vec{u}_h,\vec{w}_h) - a_1(\vec{v}_h;\vec{v}_h,\vec{w}_h).$$

By (2.7), this becomes :

$$E_h = \nu|\vec{w}_h|^2_{1,\Omega} + a_1(\vec{w}_h;\vec{u}_h,\vec{w}_h).$$

Therefore,

$$E_h \geqslant \nu|\vec{w}_h|^2_{1,\Omega} - N_h|\vec{u}_h|_{1,\Omega}|\vec{w}_h|^2_{1,\Omega} .$$

But it follows from (3.4) that

$$|\vec{u}_h|_{1,\Omega} \leqslant \frac{1}{\nu} \|\vec{f}\|^{\star}_h .$$

Hence

$$E_h \geqslant \nu\Big(1 - \frac{N_h}{\nu^2}\|\vec{f}\|^{\star}_h\Big)|\vec{w}_h|^2_{1,\Omega} ,$$

and in view of (3.16)

$$E_h \geqslant \nu\frac{\delta}{2}|\vec{w}_h|^2_{1,\Omega} . \tag{3.17}$$

Now, let us find an upper bound for E_h. By virtue of (3.4), we have :

$$E_h = \langle\vec{f},\vec{w}_h\rangle - a(\vec{v}_h;\vec{v}_h,\vec{w}_h),$$

and owing to (2.10), we can write :

$$E_h = a(\vec{u};\vec{u},\vec{w}_h) - (p,\operatorname{div}\vec{w}_h) - a(\vec{v}_h;\vec{v}_h,\vec{w}_h).$$

But since $\vec{w}_h \in V_h$, this can also be written as follows :

$$E_h = a_0(\vec{u}-\vec{v}_h,\vec{w}_h) + a_1(\vec{u};\vec{u},\vec{w}_h) - a_1(\vec{v}_h;\vec{v}_h,\vec{w}_h) - (p-q_h,\operatorname{div}\vec{w}_h) \quad \forall q_h \in M_h,$$

or equivalently,

$$E_h = a_0(\vec{u}-\vec{v}_h,\vec{w}_h) + a_1(\vec{u};\vec{u}-\vec{v}_h,\vec{w}_h) + a_1(\vec{u}-\vec{v}_h;\vec{v}_h,\vec{w}_h) - (p-q_h,\operatorname{div}\vec{w}_h).$$

Then, (2.6) yields the upper bound :

$$|E_h| \leqslant \{|\vec{u}-\vec{v}_h|_{1,\Omega}(\nu + C_1|\vec{u}|_{1,\Omega} + C_1|\vec{v}_h|_{1,\Omega}) + \|p-q_h\|_{0,\Omega}\}|\vec{w}_h|_{1,\Omega} . \tag{3.18}$$

Combining the bounds (3.17) and (3.18), we get :

$$|\vec{u}_h-\vec{v}_h|_{1,\Omega} \leq \frac{2}{\nu\delta}\{|\vec{u}-\vec{v}_h|_{1,\Omega}(\nu + C_1|\vec{u}|_{1,\Omega} + C_1|\vec{v}_h|_{1,\Omega}) + \|p-q_h\|_{0,\Omega}\}$$

$$\forall \vec{v}_h \in V_h, \quad \forall q_h \in M_h.$$

This implies :

$$|\vec{u}-\vec{u}_h|_{1,\Omega} \leq C_2\{\inf_{\vec{v}_h\in V_h} |\vec{u}-\vec{v}_h|_{1,\Omega} + \inf_{q_h\in M_h} \|p-q_h\|_{0,\Omega}\}, \tag{3.19}$$

where C_2 depends upon ν, δ and $\vec{u}$ but not upon h. The theorem follows from (3.19) and Hypotheses H1 and H2. ∎

3.2. THE NON-UNIQUENESS CASE

In this section, we assume that the Navier-Stokes problem has a nonsingular solution $\vec{u}_0 \in V$, i.e. that satisfies :

$$\sup_{\vec{w}\in V} \frac{|c(\vec{u}_0;\vec{v},\vec{w})|}{|\vec{w}|_{1,\Omega}} \geq \gamma|\vec{v}|_{1,\Omega} \quad \forall \vec{v} \in V, \tag{2.12}$$

where c is defined by :

$$c(\vec{u};\vec{v},\vec{w}) = a_0(\vec{v},\vec{w}) + a_1(\vec{u};\vec{v},\vec{w}) + a_1(\vec{v};\vec{u},\vec{w}). \tag{2.11}$$

In this case, we intend to show that problem (P_h) has a unique solution $\vec{u}_h$ in some neighborhood of the projection $\Pi_h\vec{u}_0$ on V_h. When $\vec{u}_0$ is sufficiently smooth, this neighborhood should yield a good error estimate. Finally, we want to establish that $\vec{u}_h$ can be efficiently computed by Newton's method.

More precisely, let $\Pi_h\vec{u} \in V_h$ be defined by :

$$a_0(\vec{u}-\Pi_h\vec{u},\vec{v}_h) = 0 \quad \forall \vec{v}_h \in V_h, \tag{3.20}$$

i.e.

$$|\vec{u}-\Pi_h\vec{u}|_{1,\Omega} = \inf_{\vec{v}_h\in V_h} |\vec{u}-\vec{v}_h|_{1,\Omega}.$$

The idea is to establish a discrete analogue of (2.12) in V_h with $\vec{u}_0$ replaced by $\Pi_h\vec{u}_0$. This is achieved with the help of the operator $K_h \in \mathcal{L}(X;V_h)$ defined as follows :

for each $\vec{u} \in X$, $K_h\vec{u}$ is the solution in V_h of the approximate Stokes problem

(3.21) $$a_0(K_h\vec{u},\vec{v}_h) = a_1(\Pi_h\vec{u}_0;\vec{u},\vec{v}_h) + a_1(\vec{u};\Pi_h\vec{u}_0,\vec{v}_h) \quad \forall\vec{v}_h \in V_h.$$

When combined with (2.11), this yields :

(3.22) $$a_0\big((I+K_h)\vec{u},\vec{v}_h\big) = c(\Pi_h\vec{u}_0;\vec{u},\vec{v}_h).$$

LEMMA 3.2.

Let Hypotheses H1 and H2 be satisfied and assume that Ω satisfies the regularity hypotheses of Lemma 2.3. Then, for $n=2$ or 3, we have :

$$\lim_{h\to o} \|K-K_h\|_{\mathcal{L}(X;X)} = 0.$$

Proof

Let $\vec{u}_0 \in V$ and $\vec{u}\in X$. As mentioned in the proof of Lemma 2.3, $K\vec{u}\in V$ satisfies :

(3.23) $$a_0(K\vec{u},\vec{v}) = (\vec{g},\vec{v}) = a_1(\vec{u}_0;\vec{u},\vec{v}) + a_1(\vec{u};\vec{u}_0,\vec{v}) \quad \forall\vec{v}\in V .$$

Likewise, according to (3.21), $K_h\vec{u}\in V_h$ satisfies :

(3.24) $$a_0(K_h\vec{u},\vec{v}_h) = (\vec{g}_h,\vec{v}_h) = a_1(\Pi_h\vec{u}_0;\vec{u},\vec{v}_h) + a_1(\vec{u};\Pi_h\vec{u}_0,\vec{v}_h) \quad \forall\vec{v}_h\in V_h,$$

i.e. $K_h\vec{u}$ is an approximation of the solution $\vec{z}\in V$ of the Stokes problem :

(3.25) $$a_0(\vec{z},\vec{v}) = (\vec{g}_h,\vec{v}) \quad \forall\vec{v}\in V .$$

It follows from (3.23) and (3.25) that

$$|K\vec{u}-\vec{z}|_{1,\Omega} \leq \frac{1}{\nu}\|\vec{g}-\vec{g}_h\|^\star$$

and

$$(\vec{g}-\vec{g}_h,\vec{v}) = a_1(\vec{u}_0-\Pi_h\vec{u}_0;\vec{u},\vec{v}) + a_1(\vec{u};\vec{u}_0-\Pi_h\vec{u}_0,\vec{v}) \quad \forall\vec{v}\in V.$$

Thus

$$\|\vec{g}-\vec{g}_h\|^\star \leq C_1|\vec{u}_0-\Pi_h\vec{u}_0|_{1,\Omega}|\vec{u}|_{1,\Omega}.$$

Hence

(3.26) $$|K\vec{u}-\vec{z}|_{1,\Omega} \leq C_2|\vec{u}_0-\Pi_h\vec{u}_0|_{1,\Omega}|\vec{u}|_{1,\Omega}.$$

On the other hand, let λ be *the* element of $L_0^2(\Omega)$ that satisfies :

(3.27) $$a_0(\vec{z},\vec{v}) - (\lambda,\operatorname{div}\vec{v}) = (\vec{g}_h,\vec{v}) \quad \forall\vec{v}\in X.$$

Then, Theorem 1.1, Chapter II may be applied to (3.27) and (3.24), yielding the error bound :

(3.28) $$|\vec{z}-K_h\vec{u}|_{1,\Omega} \leq C_3\{\inf_{\vec{v}_h\in V_h} |\vec{z}-\vec{v}_h|_{1,\Omega} + \inf_{q_h\in M_h} \|\lambda-q_h\|_{0,\Omega}\}.$$

Now, from (3.20) and Hypothesis H1 we infer immediately that

(3.29) $$\lim_{h\to o} |\vec{u}_0-\Pi_h\vec{u}_0|_{1,\Omega} = 0.$$

Next, like in Lemma 2.3, we find that, for $n=2$ or 3, g_h belongs to $(L^{\frac{3}{2}}(\Omega))^n$. Therefore, Theorem 5.2, Chapter I, implies that $\vec{z}\in (W^{2,\frac{3}{2}}(\Omega))^n$, $\lambda\in W^{1,\frac{3}{2}}(\Omega)$ and

$$\|\vec{z}\|_{2,\frac{3}{2},\Omega} + \|\lambda\|_{1,\frac{3}{2},\Omega} \leq C_4|\Pi_h\vec{u}_0|_{1,\Omega}|\vec{u}|_{1,\Omega} \leq C_4|\vec{u}_0|_{1,\Omega}|\vec{u}|_{1,\Omega}.$$

Furthermore, it follows from Theorem 2.2 that $W^{2,\frac{3}{2}}(\Omega)\subset H^{\frac{3}{2}}(\Omega)$ and $W^{1,\frac{3}{2}}(\Omega)\subset H^{\frac{1}{2}}(\Omega)$. Hence

(3.30) $$\|\vec{z}\|_{\frac{3}{2},\Omega} + \|\lambda\|_{\frac{1}{2},\Omega} \leq C_5|\vec{u}_0|_{1,\Omega}|\vec{u}|_{1,\Omega}.$$

Then we can use Theorem 2.3 on the interpolation in Sobolev spaces in order to derive an appropriate estimate for the right-hand side of (3.28). Indeed, Hypothesis H2 with $m=0$ and 1 gives respectively :

$$\|q-\rho_hq\|_{0,\Omega} \leq \|q\|_{0,\Omega}, \quad \|q-\rho_hq\|_{0,\Omega} \leq C_6h\|q\|_{1,\Omega}.$$

Thus the operator $I-\rho_h$ belongs to $\mathcal{L}(M;M)\cap\mathcal{L}(M\cap H^1(\Omega);M)$. Therefore, according to Theorem 2.3. $I-\rho_h\in\mathcal{L}(M\cap H^{\frac{1}{2}}(\Omega);M)$ and

(3.31) $$\|q-\rho_hq\|_{0,\Omega} \leq C_7h^{\frac{1}{2}}\|q\|_{\frac{1}{2},\Omega} \quad \forall q\in M\cap H^{\frac{1}{2}}(\Omega).$$

Likewise, (3.20) and Hypothesis H1 with $m=1$ imply :

$$|\vec{v}-\Pi_h\vec{v}|_{1,\Omega} \leq |\vec{v}|_{1,\Omega} \quad \forall\vec{v}\in V,$$

$$|\vec{v}-\Pi_h\vec{v}|_{1,\Omega} \leq |\vec{v}-r_h\vec{v}|_{1,\Omega} \leq C_8h|\vec{v}|_{2,\Omega} \quad \forall\vec{v}\in V\cap(H^2(\Omega))^n.$$

Hence

(3.32) $$|\vec{v}-\Pi_h\vec{v}|_{1,\Omega} \leq C_9h^{\frac{1}{2}}\|\vec{v}\|_{\frac{3}{2},\Omega} \quad \forall\vec{v}\in V\cap(H^{\frac{3}{2}}(\Omega))^n.$$

Finally, by piecing together (3.31), (3.32), (3.30) and (3.28), we obtain :

(3.33) $$|\vec{z}-K_h\vec{u}|_{1,\Omega} \leq C_{10}h^{\frac{1}{2}}|\vec{u}_0|_{1,\Omega}|\vec{u}|_{1,\Omega}.$$

Then (3.26) and (3.33) yield

$$|K\vec{u}-K_h\vec{u}|_{1,\Omega} \leqslant C_{11}\left(|\vec{u}_0-\Pi_h\vec{u}_0|_{1,\Omega} + h^{\frac{1}{2}}|\vec{u}_0|_{1,\Omega}\right)|\vec{u}|_{1,\Omega} \ ;$$

in view of (3.29), this proves the lemma. ∎

LEMMA 3.3.

Under the hypotheses of Lemma 3.2, there exist two constants $h_0 > 0$ and $\gamma_\star > 0$, both independent of h, such that, $\forall h \leqslant h_0$:

$$\sup_{\vec{w}_h \in V_h} \frac{c(\Pi_h\vec{u}_0;\vec{v}_h,\vec{w}_h)}{|\vec{w}_h|_{1,\Omega}} \geqslant \gamma_\star |\vec{v}_h|_{1,\Omega} \quad \forall \vec{v}_h \in V_h. \tag{3.34}$$

Proof

By virtue of Lemma 2.4, we can write :

$$I + K_h = (I + K)[I + (I+K)^{-1}(K_h - K)] \ .$$

But according to Lemma 3.2, there exists $h_0 > 0$ such that, $\forall h \leqslant h_0$:

$$\| (I + K)^{-1}(K_h - K) \|_{\mathcal{L}(X;X)} < 1.$$

Hence $I+K_h$ is nonsingular in $\mathcal{L}(X;X)$ and

$$\| (I + K_h)^{-1} \|_{\mathcal{L}(X;X)} \leqslant C. \tag{3.35}$$

Then, in view of (3.22), we can write :

$$\sup_{\vec{v}_h \in V_h} \frac{c(\Pi_h\vec{u}_0;\vec{v}_h,\vec{w}_h)}{|\vec{w}_h|_{1,\Omega}} \geqslant \frac{a_0\left((I + K_h)\vec{v}_h, (I + K_h)\vec{v}_h\right)}{|(I + K_h)\vec{v}_h|_{1,\Omega}} = \nu |(I + K_h)\vec{v}_h|_{1,\Omega}$$

$$\geqslant \frac{\nu}{C}|\vec{v}_h|_{1,\Omega} \quad ,$$

owing to (3.35). This proves (3.34) with $\gamma_\star = \frac{\nu}{C}$. ∎

Remark 3.1.

The statement of Lemma 3.3 is also valid "near" $\Pi_h\vec{u}_0$. More precisely, there exist two constants $\bar{\gamma} > 0$ and $\bar{\delta} > 0$ such that

$$\sup_{\vec{w}_h \in V_h} \frac{c(\vec{z}_h;\vec{v}_h,\vec{w}_h)}{|\vec{w}_h|_{1,\Omega}} \geqslant \bar{\gamma}|\vec{v}_h|_{1,\Omega} \quad \forall \vec{v}_h \in V_h, \ \ \forall h \leqslant h_0, \tag{3.34'}$$

for all $\vec{z}_h \in V_h$ that satisfy $|\vec{z}_h - \Pi_h \vec{u}_0|_{1,\Omega} \leq \bar{\delta}$.

Indeed, from the equation :

$$c(\vec{z}_h;\vec{v}_h,\vec{w}_h) = c(\Pi_h\vec{u}_0;\vec{v}_h,\vec{w}_h) + a_1(\vec{z}_h - \Pi_h\vec{u}_0;\vec{v}_h,\vec{w}_h) + a_1(\vec{v}_h;\vec{z}_h - \Pi_h\vec{u}_0,\vec{w}_h)$$

and Lemma 3.3, we get

$$\sup_{\vec{w}_h \in V_h} \frac{c(\vec{z}_h;\vec{v}_h,\vec{w}_h)}{|\vec{w}_h|_{1,\Omega}} \geq \left(\gamma_\star - 2N_h|\vec{z}_h - \Pi_h\vec{u}_0|_{1,\Omega}\right)|\vec{v}_h|_{1,\Omega} \quad \forall \vec{v}_h \in V_h. \qquad \blacksquare$$

Now, we are in a position to establish the existence of a solution $\vec{u}_h$ of problem (P_h) in a neighborhood of $\Pi_h\vec{u}_0$.

LEMMA 3.4

Under the hypotheses of Lemma 3.2, problem (P_h) has at least one solution $\vec{u}_h$ in the ball $B_h \subset V_h$ with center $\Pi_h\vec{u}_0$ and radius $\mathcal{R}_1(h)$, where $\mathcal{R}_1(h)$ is a constant such that

$$(3.36) \qquad 0 < \mathcal{R}_1(h) \leq C\{|\vec{u}_0 - \Pi_h\vec{u}_0|_{1,\Omega} + \|p - \rho_h p\|_{0,\Omega}\}$$

Proof

Let $T_h : V_h \mapsto V_h$ be the mapping which associates with each $\vec{v}_h \in V_h$ the element $\vec{\phi}_h = T_h\vec{v}_h$ defined by :

$$(3.37) \qquad c(\Pi_h\vec{u}_0;\vec{\phi}_h,\vec{w}_h) = a_1(\Pi_h\vec{u}_0;\vec{v}_h,\vec{w}_h) + a_1(\vec{v}_h;\Pi_h\vec{u}_0,\vec{w}_h) - a_1(\vec{v}_h;\vec{v}_h,\vec{w}_h) + \langle\vec{f},\vec{w}_h\rangle \quad \forall \vec{w}_h \in V_h.$$

Clearly, owing to Lemma 3.3, the mapping T_h is well defined by (3.37). We propose to establish that T_h has at least one fixed point in the ball B_h centered on $\Pi_h\vec{u}_0$ and with radius $\mathcal{R}_1(h)$. It follows from (3.37) that each of these fixed points is a solution of problem (P_h).

From (3.37) and (2.11), we get :

$$c(\Pi_h\vec{u}_0;\vec{\phi}_h - \Pi_h\vec{u}_0,\vec{w}_h) = a_1(\Pi_h\vec{u}_0;\vec{v}_h,\vec{w}_h) + a_1(\vec{v}_h;\Pi_h\vec{u}_0,\vec{w}_h) - a_1(\vec{v}_h;\vec{v}_h,\vec{w}_h)$$
$$+ \langle\vec{f},\vec{w}_h\rangle - a_0(\Pi_h\vec{u}_0,\vec{w}_h) - 2a_1(\Pi_h\vec{u}_0;\Pi_h\vec{u}_0,\vec{w}_h).$$

With (2.10), this becomes

$$c(\Pi_h\vec{u}_0;\vec{\phi}_h-\Pi_h\vec{u}_0,\vec{w}_h) = a_0(\vec{u}_0-\Pi_h\vec{u}_0,\vec{w}_h) + a_1(\vec{u}_0;\vec{u}_0,\vec{w}_h) - a_1(\Pi_h\vec{u}_0;\Pi_h\vec{u}_0,\vec{w}_h)$$

$$+ a_1(\Pi_h\vec{u}_0;\vec{v}_h,\vec{w}_h) + a_1(\vec{v}_h;\Pi_h\vec{u}_0,\vec{w}_h) - a_1(\vec{v}_h;\vec{v}_h,\vec{w}_h)$$

$$- a_1(\Pi_h\vec{u}_0;\Pi_h\vec{u}_0,\vec{w}_h) - (p,\operatorname{div}\vec{w}_h).$$

Hence,

$$(3.38)\quad \begin{cases} c(\Pi_h\vec{u}_0;\vec{\phi}_h-\Pi_h\vec{u}_0,\vec{w}_h) = a_0(\vec{u}_0-\Pi_h\vec{u}_0,\vec{w}_h) + a_1(\vec{u}_0-\Pi_h\vec{u}_0;\vec{u}_0,\vec{w}_h) \\ \quad + a_1(\Pi_h\vec{u}_0;\vec{u}_0-\Pi_h\vec{u}_0,\vec{w}_h) - a_1(\Pi_h\vec{u}_0-\vec{v}_h;\Pi_h\vec{u}_0-\vec{v}_h,\vec{w}_h) \\ \quad - (p-q_h,\operatorname{div}\vec{w}_h) \qquad \forall q_h\in M_h,\ \forall\vec{w}_h\in V_h. \end{cases}$$

On one hand, by virtue of Lemma 3.3, we can take $\vec{w}_h$ in V_h such that

$$|\vec{w}_h|_{1,\Omega} = 1,\ c(\Pi_h\vec{u}_0;\vec{\phi}_h-\Pi_h\vec{u}_0,\vec{w}_h) \geq \gamma_\star|\vec{\phi}_h-\Pi_h\vec{u}_0|_{1,\Omega}.$$

On the other hand, for this $\vec{w}_h$ (3.38) gives the upper bound,

$$|c(\Pi_h\vec{u}_0;\phi_h-\Pi_h\vec{u}_0,\vec{w}_h)| \leq (\nu+2C_1|\vec{u}_0|_{1,\Omega})|\vec{u}_0-\Pi_h\vec{u}_0|_{1,\Omega} + C_1|\Pi_h\vec{u}_0-\vec{v}_h|^2_{1,\Omega}$$

$$+ \|p-q_h\|_{0,\Omega} \qquad \forall q_h\in M_h.$$

Therefore,

$$|\vec{\phi}_h-\Pi_h\vec{u}_0|_{1,\Omega} \leq \varepsilon(h) + \lambda|\Pi_h\vec{u}_0-\vec{v}_h|^2_{1,\Omega},$$

where

$$\varepsilon(h) \leq C_2\{|\vec{u}_0-\Pi_h\vec{u}_0|_{1,\Omega} + \|p-\rho_h p\|_{0,\Omega}\},\qquad \lambda = C_1/\gamma_\star .$$

Hence T_h maps into itself every ball with center $\Pi_h\vec{u}_0$ and radius $\mathcal{R}$, for all positive numbers $\mathcal{R}$ that satisfy :

$$\mathcal{R} \geq \varepsilon(h) + \lambda\mathcal{R}^2,$$

i.e. for all $\mathcal{R}$ such that $\mathcal{R}_1(h) \leq \mathcal{R} \leq \mathcal{R}_2(h)$, where

$$\left.\begin{aligned} \mathcal{R}_1(h) &= \frac{1-\sqrt{1-4\lambda\varepsilon(h)}}{2\lambda} \sim \varepsilon(h) \\ &\text{and} \\ \mathcal{R}_2(h) &= \frac{1+\sqrt{1-4\lambda\varepsilon(h)}}{2\lambda} \sim \frac{1}{\lambda} - \varepsilon(h) \end{aligned}\right\}\text{when } h\to 0.$$

Moreover, it can be readily shown that T_h is continuous in each of these balls. In particular, by choosing $\mathcal{R} = \mathcal{R}_1(h)$, we find that T_h maps continously B_h into itself. Therefore Theorem 1.1 implies that T_h has at least one fixed point in B_h.∎

Now, Newton's method applied to problem (P_h) consists in finding a sequence $(\vec{u}_h^m)$ of elements of V_h defined $\forall m \geq 1$ by

$$(3.39) \qquad c(\vec{u}_h^{m-1};\vec{u}_h^m,\vec{v}_h) = a_1(\vec{u}_h^{m-1};\vec{u}_h^{m-1},\vec{v}_h) + \langle \vec{f},\vec{v}_h \rangle \quad \forall \vec{v}_h \in V_h,$$

starting from an arbitrary $\vec{u}_h^0$ in V_h. The next lemma establishes the existence and convergence of this sequence.

<u>LEMMA 3.5</u>

<u>Under the hypotheses of Lemma 3.2, there exists a constant</u> $\rho > 0$ <u>such that for all sufficiently small</u> h, <u>the condition</u>

$$(3.40) \qquad |\vec{u}_h^0 - \Pi_h \vec{u}_0|_{1,\Omega} \leq \frac{\rho}{2}$$

<u>implies that the sequence</u> $(\vec{u}_h^m)$ <u>is well defined by (3.39) and</u>

$$(3.41) \qquad |\vec{u}_h^{m+1} - \vec{u}_h|_{1,\Omega} \leq \rho, \quad |\vec{u}_h^{m+1} - \vec{u}_h|_{1,\Omega} \leq \frac{N_h}{\bar{\gamma}} |\vec{u}_h^m - \vec{u}_h|_{1,\Omega}^2 \qquad \forall m \geq 0,$$

<u>where</u> $\bar{\gamma}$ <u>is the constant of Remark 3.1.</u>

<u>Proof</u>.

Let $\bar{\delta}$ and $\bar{\gamma}$ be the constants of Remark 3.1; we set

$$\rho = \min\left(\frac{2}{3}\bar{\delta}, \frac{\bar{\gamma}}{N_h}\right).$$

Suppose that h is small enough so that $\mathcal{R}_1(h) \leq \frac{\rho}{2}$. Then, according to Lemma 3.4, problem (P_h) has at least one solution $\vec{u}_h$ such that

$$(3.42) \qquad |\vec{u}_h - \Pi_h \vec{u}_0|_{1,\Omega} \leq \frac{\rho}{2} \ .$$

Now, let $\vec{u}_h^0$ be an arbitrary element of V_h that satisfies (3.40) and assume for the moment that $\vec{u}_h^m$ is well defined by (3.39) and such that

$$(3.43) \qquad |\vec{u}_h^m - \vec{u}_h|_{1,\Omega} \leq \rho.$$

Then,

$$|\vec{u}_h^m - \Pi_h \vec{u}_0|_{1,\Omega} \leq \frac{3}{2}\rho \leq \bar{\delta} \ .$$

Therefore, Remark 3.1 implies that $\vec{u}_h^{m+1}$ is well defined by (3.39). Besides that, by virtue of (3.39), (3.4) and (2.11), we have the equality :

$$c(\vec{u}_h^m;\vec{u}_h^{m+1} - \vec{u}_h,\vec{v}_h) = a_1(\vec{u}_h^m - \vec{u}_h;\vec{u}_h^m - \vec{u}_h,\vec{v}_h).$$

Hence, (3.34') yields :

$$|\vec{u}_h^{m+1}-\vec{u}_h|_{1,\Omega} \leqslant \frac{N_h}{\bar{\gamma}} |\vec{u}_h^m-\vec{u}_h|_{1,\Omega}^2 .$$

With the induction hypothesis (3.43), this becomes

$$|\vec{u}_h^{m+1}-\vec{u}_h|_{1,\Omega} \leqslant \frac{N_h}{\bar{\gamma}} \rho^2 \leqslant \frac{N_h}{\bar{\gamma}} \rho \frac{\bar{\gamma}}{N_h} = \rho .$$

As $|\vec{u}_h^0-\vec{u}_h|_{1,\Omega}$ also satisfies (3.43), the lemma follows by induction. ■

Of course, ρ can be chosen so that $\rho < \frac{\bar{\gamma}}{N_h}$ and therefore

$$\lim_{m\to\infty} \vec{u}_h^m = \vec{u}_h .$$

In other words, if the first approximation $\vec{u}_h^0$ is sufficiently near to $\Pi_h\vec{u}_0$ then the sequence $(\vec{u}_h^m)$ converges quadratically toward $\vec{u}_h$. Furthermore, since ρ and $\vec{u}_h^0$ are independent of the particular solution $\vec{u}_h$, it follows that problem (P_h) has exactly one solution $\vec{u}_h$ in the ball with radius ρ and center $\Pi_h\vec{u}_0$. These results are summed up in the following theorem.

THEOREM 3.3

Suppose that Ω satisfies the regularity assumptions of Lemma 2.3 and that Hypotheses H1 and H2 are valid. Then, if $\vec{u}_0$ is a nonsingular solution of the Navier-Stokes problem, there exists two constants $h_0 > 0$ and $\rho > 0$ such that for all $h \in (0,h_0]$ problem (P_h) has a unique solution $\vec{u}_h \in V_h$ in the ball with radius ρ and center $\Pi_h\vec{u}_0$. Moreover, if $(\vec{u}_0,p) \in \left(V \cap (H^{m+1}(\Omega))^n\right) \times \left(L_0^2(\Omega) \cap H^m(\Omega)\right)$, we have the error bound :

$$|\vec{u}_0-\vec{u}_h|_{1,\Omega} \leqslant C\, h^m \{ \|\vec{u}_0\|_{m+1,\Omega} + \|p\|_{m,\Omega} \} \text{ for } 1 \leqslant m \leqslant \ell .$$

Furthermore, if $\vec{u}_h^0 \in V_h$ satisfies

$$|\vec{u}_h^0-\Pi_h\vec{u}_0|_{1,\Omega} \leqslant \frac{\rho}{2} ,$$

then the sequence $(\vec{u}_h^k)$ defined by Newton's method :

$$c(\vec{u}_h^{k-1};\vec{u}_h^k,\vec{v}_h) = a_1(\vec{u}_h^{k-1};\vec{u}_h^{k-1},\vec{v}_h) + \langle \vec{f},\vec{v}_h \rangle \quad \forall v_h \in V_h$$

converges quadratically toward $\vec{u}_h$.

Remark 3.2

We can apply here the arguments developed for the Stokes problem in order to derive error estimates for the pressure. ∎

§4. A MIXED METHOD FOR APPROXIMATING THE NAVIER-STOKES PROBLEM

The object of this paragraph is to adapt to the Navier-Stokes problem the mixed finite element method developed in Chapter III. The discussion is restricted to the two-dimensional case and to homogeneous boundary conditions. And of course, it is assumed that Ω is a bounded domain of $\mathbb{R}^2$ with a Lipschitz continuous boundary Γ.

4.1. A MIXED FORMULATION

Recall that the Navier-Stokes problem can be stated in terms of stream functions as follows :

$$(4.1)\quad \begin{cases} \text{Find } \psi\in\Phi \text{ (cf. Section 2.2) such that :} \\ \nu(\Delta\psi,\Delta\phi) + \int_\Omega \Delta\psi\Big(\frac{\partial\psi}{\partial x_2}\frac{\partial\phi}{\partial x_1} - \frac{\partial\psi}{\partial x_1}\frac{\partial\phi}{\partial x_2}\Big)dx = (\vec{f},\overrightarrow{\mathrm{curl}}\,\phi) \quad \forall\phi\in\Phi. \end{cases}$$

This suggests the following choice of spaces and forms :

$$X = \Phi\times L^2(\Omega),\ M = L^2(\Omega)\ ,$$

$$(4.2)\qquad V = \{v = (\phi,\theta)\in X\ ;\ \theta = -\Delta\phi\},$$

$$(4.3)\qquad a_0(u,v) = \nu(\omega,\theta) \quad \forall u = (\psi,\omega),\ v = (\phi,\theta),$$

$$(4.4)\qquad a_1(w;u,v) = \int_\Omega \tau\Big(\frac{\partial\psi}{\partial x_1}\frac{\partial\phi}{\partial x_2} - \frac{\partial\psi}{\partial x_2}\frac{\partial\phi}{\partial x_1}\Big)dx \quad \forall w = (\chi,\tau)\in X,$$

$$a(w;u,v) = a_0(u,v) + a_1(w;u,v).$$

According to (4.2), we define the form $b(\cdot,\cdot)$ as follows :

$$(4.5)\qquad b(v,\mu) = -(\Delta\phi+\theta,\mu) \quad \forall v = (\phi,\theta)\in X,\ \forall\mu\in M,$$

so that

$$V = \{v\in X\ ;\ b(v,\mu) = 0 \quad \forall\mu\in M\}.$$

Finally, we set,

$$<\ell,v> = <\vec{f},\overrightarrow{\text{curl}}\,\phi> \quad \forall v \in X. \tag{4.6}$$

As mentioned in Section 3.3, Chapter I, $\|\Delta\phi\|_{0,\Omega}$ and $\|\phi\|_{2,\Omega}$ are two equivalent norms on Φ. Hence the mapping $v = (\phi,\theta) \mapsto \|\theta\|_{0,\Omega}$ is a norm on V, equivalent to the product norm. Therefore, it is natural to define the norm of a_1 on V^3 and of ℓ on V' by :

$$N = \sup_{u,v,w \in V} \frac{|a_1(w;u,v)|}{\|\tau\|_{0,\Omega}\,\|\omega\|_{0,\Omega}\,\|\theta\|_{0,\Omega}}, \tag{4.7}$$

$$\|\ell\|^* = \sup_{v \in V} \frac{|<\ell,v>|}{\|\theta\|_{0,\Omega}}. \tag{4.8}$$

With the above notation, problem (4.1) becomes :

(P) Find $u = (\psi,\omega) \in V$ such that

$$a(u;u,v) = <\ell,v> \quad \forall v \in V. \tag{4.9}$$

Then, problem (Q) associated with (P) is :

(Q) Find a pair $(u,\lambda) \in X \times M$ satisfying

$$a(u;u,v) + b(v,\lambda) = <\ell,v> \quad \forall v \in X, \tag{4.10}$$

$$b(u,\mu) = 0 \quad \forall \mu \in M. \tag{4.11}$$

THEOREM 4.1

1°/ *Problem (P) has at least one solution* $u = (\psi,\omega)$ *in* V. *Moreover, if*

$$\frac{N}{\nu^2}\|\ell\|^* < 1, \tag{4.12}$$

this solution is unique.

2°/ *For each solution* u *of (P) there exists a unique element* λ *in* M *so that the pair* (u,λ) *satisfies problem (Q) and* $\lambda = \nu\omega$.

Proof

According to Section 2.2, problem (P) is just another formulation of the Navier-Stokes problem. Therefore part 1 follows immediately from Theorems 2.4 and 2.5.

The proof of part 2 is exactly the same as in the linear case. ■

Just like in the linear case, we propose to relax the regularity of the functions of space V in order to avoid constructing finite-dimensional subspaces of $H^2(\Omega)$. For this, we set :

$$\tilde{X} = [\tilde{\Phi} \cap W^{1,4}(\Omega)] \times L^2(\Omega), \quad \tilde{M} = H^1(\Omega)$$

and provide $\tilde{X}$ with the norm :

$$\|v\|_{\tilde{X}} = |\phi|_{1,4,\Omega} + \|\theta\|_{0,\Omega} \quad \forall v = (\phi,\theta) \in \tilde{X}.$$

Note that by Sobolev's imbedding Theorem 2.1, the form $a_1(w;u,v)$ is defined and continuous on $\tilde{X}^3$. Moreover, it satisfies trivially the equality

$$(4.13) \qquad a_1(w;v,v) = 0 \quad \forall w,v \in \tilde{X}.$$

Next, we define :

$$(4.14) \qquad \tilde{b}(v,\mu) = (\overrightarrow{\mathrm{curl}}\,\phi, \overrightarrow{\mathrm{curl}}\,\mu) - (\theta,\mu) \quad \forall v = (\phi,\theta) \in \tilde{X}, \quad \forall \mu \in \tilde{M},$$

$$(4.15) \qquad \tilde{V} = \{v \in \tilde{X} \; ; \; \tilde{b}(v,\mu) = 0 \quad \forall \mu \in \tilde{M}\},$$

and we recall that according to Lemma 2.1, Chapter III,

$$\tilde{V} = V.$$

Therefore problem $(\tilde{P})$ coincides with problem (P), and problem $(\tilde{Q})$ associated with the form $\tilde{b}$ and the space $\tilde{X}$ is :

$$(\tilde{Q}) \begin{cases} \text{Find } u \in \tilde{V} \text{ and } \lambda \in \tilde{M} \text{ satisfying} \\ (4.16) \qquad a(u;u,v) + \tilde{b}(v,\lambda) = \langle \ell, v \rangle \quad \forall v \in \tilde{X}. \end{cases}$$

As in the linear case, we have the following equivalence theorem with a similar proof.

<u>THEOREM 4.2</u>

<u>The set of solutions</u> (u,λ) <u>of problem</u> $(\tilde{Q})$ <u>coincideswith the set</u> (u,λ) <u>of solutions of problem (Q) such that</u> $\lambda \in H^1(\Omega)$.

4.2. AN ABSTRACT MIXED APPROXIMATION

Let us discretize the weak formulations $(\tilde{P})$ and $(\tilde{Q})$. We introduce finite dimensional spaces Φ_h, Θ_h and M_h such that

$$\Phi_h \subset \tilde{\Phi} \cap W^{1,4}(\Omega),\ \Theta_h \subset L^2(\Omega),\ M_h \subset \tilde{M}$$

and we assume that $\Phi_h \subset M_h$. Then, we set

$$X_h = \Phi_h \times \Theta_h \subset \tilde{X},$$

$$V_h = \{v_h \in X_h\ ;\ \tilde{b}(v_h,\mu_h) = 0 \quad \forall \mu_h \in M_h\}. \tag{4.17}$$

According to Lemma 2.2 Chapter III, the mapping $v_h \mapsto \|\theta_h\|_{0,\Omega}$ is a norm on V_h uniformly equivalent to the norm

$$\|v_h\| = \{|\phi_h|^2_{1,\Omega} + \|\theta_h\|^2_{0,\Omega}\}^{\frac{1}{2}}.$$

We set

$$|v_h| = \|\theta_h\|_{0,\Omega}.$$

Furthermore, since the dimension of V_h is finite, $|v_h|$ is also equivalent (with an eventual dependence upon h) to $\|v_h\|_{\tilde{X}}$. Therefore, a_1 is continuous on V_h^3 with respect to $|\cdot|$, and we define its norm by :

$$N_h = \sup_{u_h,v_h,w_h \in V_h} \frac{|a_1(w_h;u_h,v_h)|}{|u_h|\,|v_h|\,|w_h|}. \tag{4.18}$$

Similarly, we set

$$\|\ell\|^{*}_h = \sup_{v_h \in V_h} \frac{<\ell,v_h>}{|v_h|}. \tag{4.19}$$

With these notations, we discretize $(\tilde{P})$ and $(\tilde{Q})$ as follows :

(P_h) Find $u_h = (\psi_h,\omega_h) \in V_h$ satisfying

$$a(u_h;u_h,v_h) = <\ell,v_h> \qquad v_h \in V_h. \tag{4.20}$$

(Q_h) Find $u_h \in V_h$ and $\lambda_h \in M_h$ such that

$$a(u_h;u_h,v_h) + \tilde{b}(v_h,\lambda_h) = <\ell,v_h> \quad \forall v_h \in X_h. \tag{4.21}$$

Using (4.13), it is easy to prove the following existence and uniqueness result.

THEOREM 4.3.

Problem (P_h) has at least one solution $u_h \in V_h$. Under the condition

(4.22) $$\frac{N_h}{\nu^2}\|\ell\|_h^* < 1,$$

this solution is unique.

In order to establish the convergence of u_h, we require some discrete analogue of the compactness property of the canonical imbedding from $H^2(\Omega)$ into $W^{1,4}(\Omega)$. Since Φ_h is not contained in $H^2(\Omega)$, we must assume the following :

Hypothesis H1

1°/ There exists a constant C independent of h such that

(4.23) $$|\phi_h|_{1,4,\Omega} \leq C\|\theta_h\|_{0,\Omega} \quad \forall v_h = (\phi_h,\theta_h) \in V_h.$$

2°/ If $v_h = (\phi_h,\theta_h) \in V_h$ is such that

$$\text{weak} \lim_{h\to o} v_h = v = (\phi,\theta) \text{ in } H^1(\Omega)\times L^2(\Omega),$$

then

$$\lim_{h\to o} \phi_h = \phi \text{ in } W^{1,4}(\Omega).$$ ■

Now we are in a position to prove the following lemma.

LEMMA 4.1

Let Hypothesis H1 be satisfied and suppose in addition that

(4.24) $$\lim_{h\to o} \inf_{v_h\in V_h} \|v-v_h\|_{\tilde{X}} = 0 \quad \forall v \in V,$$

(4.25) $$\lim_{h\to o} \inf_{\mu_h\in M_h} \|\mu-\mu_h\|_{1,\Omega} = 0 \quad \forall \mu \in \tilde{M}.$$

Then

$$\lim_{h\to o} N_h = N \text{ and } \lim_{h\to o} \|\ell\|_h^* = \|\ell\|^*.$$

Proof

Let us briefly establish the first limit since the proof is much the same as that of Lemma 3.1. Let u_h, v_h and w_h be three normalized elements of V_h that realize the maximum in (4.18), i.e.

$$N_h = |a_1(w_h;u_h,v_h)|$$

and $|u_h| = |v_h| = |w_h| = 1$. Then from Hölder's inequality and (4.23), it follows that

$$|a_1(w_h;u_h,v_h)| \leq \sqrt{2}\, C^2,$$

and from the equivalence of norms, it follows that $\|u_h\|$, $\|v_h\|$ and $\|w_h\|$ are bounded. Therefore, we can extract a subsequence h_p from h such that

$$\text{weak} \lim_{p\to\infty} u_{h_p} = u_\star, \text{ etc...} \text{ in } H^1(\Omega) \times L^2(\Omega),$$

$$\lim_{p\to\infty} N_{h_p} = N_\star .$$

Now, $u_\star \in V$ thanks to (4.25) and the fact that $\tilde{\Phi}$ is a closed subspace of $H^1(\Omega)$. Besides that, Hypothesis H1 n°2 implies that

$$\lim_{p\to\infty} \psi_{h_p} = \psi_\star \text{ etc... in } W^{1,4}(\Omega).$$

Therefore

$$\lim_{p\to\infty} a_1(w_{h_p};u_{h_p},v_{h_p}) = a_1(w_\star;u_\star,v_\star).$$

Hence

$$N_\star = |a_1(w_\star;u_\star,v_\star)|.$$

But since $|u_\star| \leq 1$, etc..., we get

$$N \geq \frac{|a_1(w_\star;u_\star,v_\star)|}{|u_\star||v_\star||w_\star|} \geq N_\star .$$

The reverse inequality is an easy consequence of (4.24). ∎

The next theorem establishes the strong convergence of u_h when (4.12) holds and its weak convergence in the general case.

THEOREM 4.4.

Suppose that Hypotheses H1, (4.24) and (4.25) are valid.

1°/ If the condition (4.12) holds and if the solution $u = (\psi,\omega)$ of problem (P) satisfies $\omega \in H^1(\Omega)$, then for all sufficiently small h problem (P_h) has a unique solution $u_h \in V_h$ and

(4.26) $$|u-u_h| \leq C\{ \inf_{v_h \in V_h} \| u-v_h \|_{\widetilde{X}} + \inf_{\mu_h \in M_h} \| \omega-\mu_h \|_{1,\Omega} \}.$$

2°/ <u>In the general case, as</u> h <u>tends to zero, the set of solutions</u> $u_h \in V_h$ <u>of problem</u> (P_h) <u>has at least one limit point in the weak topology of</u> $H^1(\Omega) \times L^2(\Omega)$. <u>Each limit point is a solution of problem</u> (P).

<u>Proof</u>

1°/ If (4.12) is valid, there exists a number $\delta > 0$ such that

(4.12') $$\frac{N}{\nu^2} \| \ell \|^* \leq 1 - \delta.$$

Then, according to Lemma 4.1, the inequality :

(4.27) $$\frac{N_h}{\nu^2} \| \ell \|_h^* \leq 1 - \frac{\delta}{2}$$

holds for all sufficiently small h. Hence, by virtue of Theorem 4.3, problem (P_h) has a unique solution u_h in V_h. Now, let v_h be an arbitrary element of V_h, let $w_h = u_h - v_h \in V_h$ and consider the expression :

$$E_h = a(u_h;u_h,w_h) - a(v_h;v_h,w_h),$$

i.e.

$$E_h = a_0(w_h,w_h) + a_1(w_h;u_h,w_h).$$

In view of (4.18), we have :

$$E_h \geq (\nu - N_h |u_h|) |w_h|^2,$$

and because of (4.13), we have :

(4.28) $$|u_h| \leq \frac{1}{\nu} \| \ell \|_h^*.$$

Therefore, it follows from (4.27) that

(4.29) $$E_h \geq \frac{\delta}{2} |w_h|^2.$$

On the other hand, (4.20) yields

$$E_h = \langle \ell, w_h \rangle - a(v_h;v_h,w_h).$$

With (4.16), this becomes

$$E_h = a(u;u,w_h) + \widetilde{b}(w_h,\lambda) - a(v_h;v_h,w_h).$$

Then by virtue of (4.17), we get $\forall \mu_h \in M_h$:

$$E_h = a_0(u-v_h, w_h) + a_1(u; u-v_h, w_h) + a_1(u-v_h; v_h, w_h) + \tilde{b}(w_h, \lambda-\mu_h).$$

Hence, by Hölder's inequality and (4.23), we find :

$$|E_h| \leq \{\nu|u-v_h| + C_1(|u||\psi-\phi_h|_{1,4,\Omega} + |u-v_h||v_h|) + C_2\|\lambda-\mu_h\|_{1,\Omega}\}|w_h|.$$

With (4.29) this yields :

$$|w_h| \leq C_3\{|u-v_h|(\nu + C_1|v_h|) + C_1|u|\,\|u-v_h\|_{\tilde{X}} + C_2\|\lambda-\mu_h\|_{1,\Omega}\}.$$

Therefore

$$|u-u_h| \leq C_4\{\inf_{v_h\in V_h}\|u-v_h\|_{\tilde{X}} + \inf_{\mu_h\in M_h}\|\lambda-\mu_h\|_{1,\Omega}\}.$$

2°/ Now, discard the hypotheses of part 1 and consider a solution u_h of problem (P_h). By virtue of (4.28) and Lemma 4.1, we have :

$$\|u_h\| \leq C_5.$$

Hence, the set of solutions of problems (P_h) remains bounded in norm $\|\cdot\|$ when h tends to zero. Therefore this set has at least one limit point in the weak topology of $H^1(\Omega) \times L^2(\Omega)$. Let $u = (\psi,\omega)$ be one of these limit points; then there exists a subsequence h_p of h and for each p a solution of problem (P_{h_p}): $u_{h_p} = (\psi_{h_p}, \omega_{h_p})$ such that $\lim_{p\to\infty} h_p = 0$ and

$$\text{weak} \lim_{p\to\infty} u_{h_p} = u \text{ in } H^1(\Omega) \times L^2(\Omega).$$

As mentionned in the proof of Lemma 4.1, $u \in V$ and according to H1,

$$\lim_{p\to\infty} \psi_{h_p} = \psi \text{ in } W^{1,4}(\Omega).$$

Therefore, by virtue of (4.24), we have :

$$\lim_{p\to\infty} a_1(u_{h_p}; u_{h_p}, \Pi_{h_p} v) = a_1(u;u,v) \quad \forall v \in V,$$

where Π_h denotes the projection operator on V_h for the norm $\|\cdot\|$. Hence

$$a(u;u,v) = \langle \ell, v\rangle \quad \forall v \in V .$$

■

Note that the statement of Theorem 4.4 is pretty weak in the general case. In fact, by using sophisticated arguments, it is possible to arrive at a similar conclusion to that of Theorem 3.3. For further details the reader can refer to Girault & Raviart [23] .

4.3. APPLICATIONS

To simplify the discussion, we assume that Ω is a bounded and convex domain of $\mathbb{R}^2$ with a polygonal boundary Γ. Then, we introduce a uniformly regular family of triangulations of $\bar{\Omega}$, $\mathcal{T}_h$ (cf. Definition 2.3, Chapter II) made of triangles whose diameters are bounded by h. We choose the same spaces as in Chapter III, namely :

$$(4.30)\quad \begin{cases} \Theta_h = M_h = \{\theta_h \in \mathcal{C}^0(\bar{\Omega});\ \theta_h|_K \in P_\ell \quad \forall K \in \mathcal{T}_h\}, \\ \text{where } \ell \geq 1 \text{ is an integer,} \\ \Phi_h = \Theta_h \cap H_0^1(\Omega) \end{cases}$$

$\left(\tilde{\Phi} = H_0^1(\Omega)\right.$ because Ω is simply connected$\left.\right)$.

Here, the major difficulty is the verification of Hypothesis H1.

LEMMA 4.2.

With the above choice of spaces and assumptions on Ω, there exists a constant C independent of h such that

$$(4.23)\qquad |\phi_h|_{1,4,\Omega} \leq C\|\theta_h\|_{0,\Omega} \qquad \forall v_h = (\phi_h,\theta_h) \in V_h.$$

Proof

When $v = (\phi,\theta) \in V$, ϕ and θ are related by $\theta = -\Delta\phi$ and v satisfies : $|\phi|_{1,4,\Omega} \leq C\|\theta\|_{0,\Omega}$ by virtue of Sobolev's imbedding Theorem 2.1. As V_h is intended to approximate V, we can reasonably expect θ_h to be a discretization of "$-\Delta\phi_h$", for $v_h = (\phi_h,\theta_h)$ in V_h. Thus, we introduce the solution $\tilde{\phi}(h)$ of the Dirichlet's problem :

$$(4.31)\qquad -\Delta\tilde{\phi}(h) = \theta_h \text{ in } \Omega,\ \tilde{\phi}(h) = 0 \text{ on } \Gamma,$$

and we compare ϕ_h and $\tilde{\phi}(h)$.

Since Ω is plane, bounded and convex, $\tilde{\phi}(h)$ belongs to $H^2(\Omega) \cap H_0^1(\Omega)$ and

$$(4.32)\qquad \|\tilde{\phi}(h)\|_{2,\Omega} \leq C_1\|\theta_h\|_{0,\Omega}.$$

Then Theorem 2.1 implies that :

$$(4.33)\qquad \|\tilde{\phi}(h)\|_{1,4,\Omega} \leq C_2\|\theta_h\|_{0,\Omega}.$$

In addition, problem (4.31) has the variational formulation

$$(\overrightarrow{\mathrm{curl}}\,\tilde{\phi}(h), \overrightarrow{\mathrm{curl}}\,\mu) = (\theta_h, \mu) \qquad \forall \mu \in H_0^1(\Omega),$$

and condition (4.17) on v_h reads :

$$(\overrightarrow{\mathrm{curl}}\,\phi_h, \overrightarrow{\mathrm{curl}}\,\mu_h) = (\theta_h, \mu_h) \qquad \forall \mu_h \in M_h \subset H^1(\Omega).$$

Therefore, the following equation holds for all $\mu_h \in M_h \cap H_0^1(\Omega)$:

(4.34) $$(\overrightarrow{\mathrm{curl}}(\tilde{\phi}(h) - \phi_h), \overrightarrow{\mathrm{curl}}\,\mu_h) = 0.$$

Let ϖ_h denote the standard interpolation operator of $\mathcal{L}(H^2(\Omega) \cap H_0^1(\Omega); \Phi_h)$. Then :

$$|\phi_h - \tilde{\phi}(h)|_{1,4,\Omega} \leq |\phi_h - \varpi_h \tilde{\phi}(h)|_{1,4,\Omega} + |\varpi_h \tilde{\phi}(h) - \tilde{\phi}(h)|_{1,4,\Omega}.$$

First, Theorem 2.4, Chapter II and (4.33) yield

(4.35) $$|\tilde{\phi}(h) - \varpi_h \tilde{\phi}(h)|_{1,4,\Omega} \leq C_3 \|\theta_h\|_{0,\Omega}.$$

Therefore, it suffices to estimate $|\phi_h - \varpi_h \tilde{\phi}(h)|_{1,4,\Omega}$. On the one hand, by the uniform regularity of $\mathcal{T}_h$, the following inverse inequality holds on Φ_h :

(4.36) $$|\phi_h - \varpi_h \tilde{\phi}(h)|_{1,4,\Omega} \leq C_4 h^{-\frac{1}{2}} |\phi_h - \varpi_h \tilde{\phi}(h)|_{1,\Omega}.$$

On the other hand, (4.34) implies that

(4.37) $$|\phi_h - \varpi_h \tilde{\phi}(h)|_{1,\Omega} \leq |\tilde{\phi}(h) - \varpi_h \tilde{\phi}(h)|_{1,\Omega}.$$

By applying again Theorem 2.4, Chapter II, we derive :

(4.38) $$|\tilde{\phi}(h) - \varpi_h \tilde{\phi}(h)|_{1,\Omega} \leq C_5 h \|\tilde{\phi}(h)\|_{2,\Omega}.$$

Therefore, combining (4.36), (4.37), (4.38) and (4.32), we get :

(4.39) $$|\phi_h - \varpi_h \tilde{\phi}(h)|_{1,4,\Omega} \leq C_6 h^{\frac{1}{2}} \|\theta_h\|_{0,\Omega}.$$

Together with (4.35) and (4.33), this gives :

$$|\phi_h|_{1,4,\Omega} \leq \{C_2 + C_3 + C_6 h^{\frac{1}{2}}\} \|\theta_h\|_{0,\Omega},$$

thus proving the lemma. ∎

<u>Remark 4.1.</u>

The above argument remains valid when $\tilde{\phi}(h)$ only belongs to $H^{\frac{3}{2}}(\Omega)$. Indeed, by

Theorem 2.2, $H^{\frac{3}{2}}(\Omega) \subset W^{1,4}(\Omega)$, therefore (4.33) still holds. Besides that, we can replace ϖ_h by the projection operator $P_h \in \mathcal{L}\left(H_0^1(\Omega);\Phi_h\right)$ defined by :

$$\left(\overrightarrow{\mathrm{curl}}(P_h\phi-\phi),\ \overrightarrow{\mathrm{curl}}\ \mu_h\right) = 0 \quad \forall \mu_h \in \Phi_h.$$

Then, it can be shown that

$$|\tilde{\phi}(h)-P_h\tilde{\phi}(h)|_{1,4,\Omega} \leq C_3' \| \theta_h \|_{0,\Omega}$$

and

$$|\tilde{\phi}(h)-P_h\tilde{\phi}(h)|_{1,\Omega} \leq C_5' h^{\frac{1}{2}} \| \tilde{\phi}(h) \|_{\frac{3}{2},\Omega} .$$

The end of the proof is unchanged.

Relaxing the regularity of $\tilde{\phi}(h)$ permits to relax the regularity assumptions on Ω. In particular, we need no longer assume that Ω is convex. ■

The next lemma checks the second part of Hypothesis H1.

<u>LEMMA 4.3.</u>

<u>Assume that the hypotheses of Lemma 4.2 hold. If a sequence</u> $v_h = (\phi_h,\theta_h) \in V_h$ <u>satisfies</u>

$$\text{weak} \lim_{h\to o} v_h = v \ \underline{\text{in}}\ H_0^1(\Omega) \times L^2(\Omega),$$

<u>then</u>

$$\lim_{h\to o} \phi_h = \phi \ \underline{\text{in}}\ W^{1,4}(\Omega).$$

<u>Proof</u>

The idea of the proof is much the same as that of Lemma 4.2. Again let $\tilde{\phi}(h)$ be the solution of problem (4.31). Then, since $\theta = -\Delta\phi$, we have

$$\left(\overrightarrow{\mathrm{curl}}(\tilde{\phi}(h)-\phi),\overrightarrow{\mathrm{curl}}\ \mu\right) = (\theta_h-\theta,\mu) \quad \forall \mu \in H_0^1(\Omega).$$

Therefore,

$$\text{weak} \lim_{h\to o} \tilde{\phi}(h) = \phi \text{ in } H_0^1(\Omega),$$

since weak $\lim_{h\to o} \theta_h = \theta$ in $L^2(\Omega)$. Furthermore, by virtue of (4.32),

$$\text{weak} \lim_{h\to o} \tilde{\phi}(h) = \phi \text{ in } H^2(\Omega).$$

Hence

$$\lim_{h\to o} \tilde{\phi}(h) = \phi \text{ in } W^{1,4}(\Omega). \tag{4.40}$$

Now let us write

$$|\phi_h-\phi|_{1,4,\Omega} \leq |\phi_h-\varpi_h\tilde{\phi}(h)|_{1,4,\Omega} + |\varpi_h(\tilde{\phi}(h)-\phi)|_{1,4,\Omega} + |\varpi_h\phi-\phi|_{1,4,\Omega} .$$

According to (4.39), we have :

$$\lim_{h\to o}\left(\phi_h-\varpi_h\tilde{\phi}(h)\right) = 0 \text{ in } W^{1,4}(\Omega).$$

Next, Theorem 2.4, Chapter II and (4.40) imply that :

$$|\varpi_h(\tilde{\phi}(h)-\phi)|_{1,4,\Omega} \leq C_1|\tilde{\phi}(h)-\phi|_{1,4,\Omega} \to 0 \text{ as } h \to 0.$$

Finally, by applying again this theorem to all ϕ in $W^{2,4}(\Omega)\cap H_0^1(\Omega)$, we get :

$$|\varpi_h\phi-\phi|_{1,4,\Omega} \leq C_2 h \|\phi\|_{2,4,\Omega}.$$

Therefore, by density

$$\lim_{h\to o}|\varpi_h\phi-\phi|_{1,4,\Omega} = 0 \quad \forall\phi\in W^{1,4}(\Omega)\cap H_0^1(\Omega).$$

These three limits imply that $\lim_{h\to o} \phi_h = \phi$ in $W^{1,4}(\Omega)$. ∎

Remark 4.2.

Here again, the regularity of $\tilde{\phi}(h)$ can be relaxed but slightly less than in Lemma 4.2. In fact, in order to derive the strong convergence of $\tilde{\phi}(h)$ in $W^{1,4}(\Omega)$, we require the weak convergence of $\tilde{\phi}(h)$ in $H^{\frac{3}{2}+\varepsilon}(\Omega)$ for any $\varepsilon>0$. ∎

Now we can establish the convergence of u_h.

THEOREM 4.5.

For $\ell\geq 2$, let Φ_h, Θ_h and M_h be defined by (4.30) on a uniformly regular family of triangulations of $\bar{\Omega}$.

1°/ When h tends to zero, the set of solutions u_h of (P_h) has at least one limit point for the weak topology of $H_0^1(\Omega)\times L^2(\Omega)$. Each one is a solution of problem (P).

2°/ Assume that (4.12) holds and that the solution u of (P) is such that $\psi\in H^{\ell+2}(\Omega)$. Then the following error estimate holds :

(4.41) $$\|u-u_h\| \leq C h^{\ell-1}\{|\psi|_{\ell+1,\Omega} + |\psi|_{\ell+2,\Omega}\}.$$

Proof.

Let us check (4.24) and (4.25). The space M_h is a classical approximation of $H^1(\Omega)$ and it is well known that

$$\lim_{h\to 0} \inf_{\mu_h \in M_h} \|\omega-\mu_h\|_{1,\Omega} = 0.$$

As far as (4.24) is concerned, we have proved in Corollary 1.1 , Chapter III that

$$\inf_{v_h \in V_h} \|u-v_h\|_{\tilde{X}} \leq C_1\big(1 + S(h)\big) \inf_{v_h \in X_h} \|u-v_h\|_{\tilde{X}},$$

where, according to Lemma 2.4, Chapter III,

$$S(h) \leq \frac{C_2}{h}.$$

Now, when $\psi \in H^4(\Omega)$ then $\omega = -\Delta\psi \in H^2(\Omega)$ and since $\ell \geq 2$, we have :

$$\inf_{v_h \in V_h} \|u-v_h\|_{\tilde{X}} \leq C_3 h \|\psi\|_{4,\Omega}.$$

Therefore by density we obtain that for any u in V:

$$\lim_{h\to 0} \inf_{v_h \in V_h} \|u-v_h\|_{\tilde{X}} = 0.$$

This proves the first part of the theorem.

The second part of the theorem follows from (4.26) and the estimates derived in the proof of Theorem 2.4, Chapter III. ∎

Remark 4.3.

As mentioned at the end of Section 4.2, the statement of Theorem 4.5 can be greatly improved. Furthermore, like in the linear case, the convergence can also be established when $\ell = 1$. ∎

CHAPTER V

THE TIME-DEPENDENT NAVIER-STOKES EQUATIONS

§ 1 - THE CONTINUOUS PROBLEM

We first introduce the theoretical material required to handle the time variable in a boundary-value problem. Then we derive an adequate variational formulation of the time-dependent Navier-Stokes equations. We then prove the existence theorem and a uniqueness result.

1.1. Some vector-valued function spaces.

When dealing with $u(x,t)$, a function of a space variable x and a time variable t, it is often convenient to separate the variables and consider u as $u(t)$, a function of time only that takes its values in a function space. That is, for each t, $u(t)$ is the mapping $x \longmapsto u(x,t)$. The function $t \longmapsto u(t)$ is usually called a vector-valued function. We propose to extend to such functions the familiar notions of spaces L^p and $\mathcal{C}^o$, the idea of derivative, etc ...

Let (a,b) be an interval of the extended line, i.e. $-\infty \leq a < b \leq +\infty$, and X a Banach space normed by $\|.\|_X$. For real $p \geq 1$, we denote by $L^p(a,b\,;X)$ the space of strongly measurable functions $f:(a,b) \longmapsto X$ (i.e. $t \longmapsto \|f(t)\|_X$ is measurable) such that

$$\|f\|_{L^p(a,b\,;\,X)} = \left[\int_a^b \|f(t)\|_X^p \, dt\right]^{1/p} < \infty \quad \text{if} \quad 1 \leq p < \infty$$

or

$$\|f\|_{L^\infty(a,b\,;\,X)} = \sup_{t \in (a,b)} \text{ess} \, \|f(t)\|_X < \infty \qquad \text{if} \quad p = \infty .$$

When $-\infty < a < b < +\infty$, we denote by $\mathcal{C}^o([a,b]\,;X)$ the space of continuous functions $f : [a,b] \longmapsto X$ normed by

$$\|f\|_{\mathcal{C}^\circ([a,b]\,;\,X)} = \max_{t \in [a,b]} \|f(t)\|_X .$$

Now, let us introduce the notion of generalized derivative of a vector-valued function.

LEMMA 1.1.

Let X be a Banach space, X' its dual space and let u and g be two functions of $L^1(a,b\,;\,X)$. The three following conditions are equivalent :

(i) for some ξ in X u satisfies :

$$u(t) = \xi + \int_a^t g(s)\,ds \quad \text{a.e. in } (a,b) \; ; \tag{1.1}$$

(ii) for all functions φ of $\mathcal{D}(]a,b[)$, we have :

$$\int_a^b u(t)\varphi'(t)\,dt = -\int_a^b g(t)\varphi(t)\,dt \; ; \tag{1.2}$$

(iii) for all η in X' , u satisfies

$$\frac{d}{dt} < u,\eta >_X = < g,\eta >_X \quad \text{in} \quad \mathcal{D}'(]a,b[) . \tag{1.3}$$

Furthermore in each of these cases, u is almost everywhere equal to a function of $\mathcal{C}^\circ([a,b]\,;\,X)$.

The proof of this lemma can be found in Temam [44].

The statement of Lemma 1.1 suggests the following definition :

DEFINITION 1.1.

The function g of the above lemma is called the weak or generalized derivative of u and is denoted by the usual symbol :

$$g = u' = \frac{du}{dt} .$$

Obviously, this definition can be extended to higher-order derivatives.

Now, we are in a position to introduce and study closely the spaces of vector-valued functions that are best adapted to the solution of time-dependen problems. At this stage, there is no need to specify these spaces and therefor we consider the following abstract situation. Let V and H be two Hilbert spaces normed respectively by $\|.\|$ and $|.|$, and such that

(1.4) V is contained and dense in H with a continuous imbedding.

Let $(.,.)$ denote the scalar product of H corresponding to $|.|$; also let be the dual space of V and, as usual, let $\|.\|_*$ denote the norm of V' . If we decide to identify H with its dual space H' by means of the scalar product $(.,.)$, then H can be identified with a subspace of V' and the follo dense and continuous inclusions hold :

$$V \subset H \subset V' . \tag{1.5}$$

Furthermore, the operation $<.,.>$ expressing the duality between V and is simply an extension of the scalar product $(.,.)$.

Let $T > 0$ be a fixed real number and consider the space

$$W(0,T) \equiv W(0,T\ ;\ V,V') = \{v \in L^2(0,T\ ;\ V)\ ;\ \frac{dv}{dt} \in L^2(0,T\ ;\ V')\} \ , \tag{1.6}$$

normed by

$$\|v\|_{W(0,T)} = \left[\int_0^T (\|v(t)\|^2 + \|\frac{dv(t)}{dt}\|_*^2)dt\right]^{1/2} . \tag{1.7}$$

Then $W(0,T)$ is a Hilbert space. According to Lemma 1.1, an element of $W(0,$ coincides almost everywhere with a function of $\mathcal{C}^\circ([0,T]\ ;\ V')$. In fact, we shall prove that $v \in \mathcal{C}^\circ([0,T]\ ;\ H)$. Let us first establish a preliminary l

<u>LEMMA 1.2.</u>

<u>The space</u> $\mathcal{C}^\infty([0,T]\ ;\ V)$ <u>is dense in</u> $W(0,T\ ;\ V,V')$.

<u>PROOF.</u>

<u>Let</u> α <u>and</u> β <u>be two scalar functions of</u> $\mathcal{C}^\infty([0,T])$ <u>such that</u>

$$\begin{cases} 0 \leq \alpha(t),\beta(t) \leq 1\ ,\quad \alpha + \beta \equiv 1 \text{ in } [0,T]\ , \\ \text{supp}(\alpha) \subset [0, \frac{2T}{3}]\ ,\quad \text{supp}(\beta) \subset [\frac{T}{3},T]\ . \end{cases} \tag{1.8}$$

Let $u \in W(0,T)$. According to (1.8), we can write :

$$u = \alpha u + \beta u .$$

Thus, it suffices to construct a sequence in $\mathcal{C}^{\infty}([0,T];V)$ that converges to αu (for instance). Let $v = \alpha u$ and let $\tilde{v}$ be the extension of v by zero beyond T ; then $\tilde{v} \in W(0,\infty)$ since α vanishes beyond $\frac{2T}{3}$. Now, for $h > 0$, consider the function v_h defined by

$$v_h(t) = \tilde{v}(t+h) \qquad \forall\, t \geq -h .$$

Clearly

$$\lim_{h \to 0} v_h = \tilde{v} \quad \text{in} \quad W(0,\infty) .$$

Therefore, it suffices to approach v_h with a sequence of $\mathcal{C}^{\infty}([0,T];V)$. This is achieved by means of a classical regularization device.

Let $\rho \in \mathcal{D}(\mathbb{R})$ with $\rho \geq 0$, $\int_{\mathbb{R}} \rho(t)dt = 1$ and $\operatorname{supp}(\rho) \subset [-1,1]$.

For any $\varepsilon > 0$, let $\rho_\varepsilon(t) = \rho(\frac{t}{\varepsilon})$. Then

$$\rho_\varepsilon \in \mathcal{D}(\mathbb{R}) \quad , \quad \lim_{\varepsilon \to 0} \rho_\varepsilon = \delta \quad \text{in} \quad \mathcal{D}'(\mathbb{R}) .$$

Therefore, $\rho_\varepsilon * v_h \in \mathcal{C}^{\infty}([0,T];V)$,

$$\lim_{\varepsilon \to 0} (\rho_\varepsilon * v_h) = v_h \quad \text{in} \quad L^2(0,\infty ; V) ,$$

$$\lim_{\varepsilon \to 0} \frac{d}{dt}(\rho_\varepsilon * v_h) = \frac{dv_h}{dt} \quad \text{in} \quad L^2(0,\infty ; V').$$

Hence $\rho_\varepsilon * v_h$ is the required sequence. ■

<u>**THEOREM 1.1.**</u>

<u>The following inclusion holds algebraically and topologically</u> :

$$W(0,T) \subset \mathcal{C}^{0}([0,T];H) . \tag{1.9}$$

<u>Moreover, the following Green's formula holds for all</u> u <u>and</u> v <u>in</u> $W(0,T)$:

$$\int_0^T \{< \frac{du}{dt}(t),v(t) > + < u(t), \frac{dv}{dt}(t) >\}dt = < u(T),v(T) > - < u(0),v(0) > \tag{1.10}$$

<u>**PROOF.**</u>

Since $\mathcal{C}^{\infty}([0,T];V)$ is dense in $W(0,T)$, it suffices to show that

(1.11) $$\max_{t \in [0,T]} |u(t)| \leq C \|u\|_{W(0,T)} \quad \forall\, u \in \mathcal{C}^\infty([0,T];V)$$

in order to obtain (1.9). Indeed, if (1.11) holds, the identity mapping from $\mathcal{C}^\infty([0,T];V)$ onto itself can be extended by continuity to a linear contin mapping of $W(0,T)$ into $\mathcal{C}^0([0,T];H)$.

Now, let $u \in \mathcal{C}^\infty([0,T];V)$ and consider the two functions α and β of $\mathcal{C}^\infty([0,T])$ that satisfy (1.8). Again we can write

$$u = v + w \quad \text{where} \quad v = \alpha u \quad \text{and} \quad w = \beta u .$$

Let us examine v. Because of the identity

$$\frac{d}{dt}|v(t)|^2 = 2\left(v(t), \frac{d}{dt}v(t)\right) \quad \forall\, t \in [0,T],$$

we have

$$|v(t)|^2 = -2\int_t^T \left(v(s), \frac{d}{ds}v(s)\right)ds \leq 2\int_0^T \|v(s)\| \|\frac{d}{ds}v(s)\|_* ds .$$

Therefore

$$|v(t)|^2 \leq \int_0^T \{\|v(s)\|^2 + \|\frac{d}{ds}v(s)\|_*^2\}\, ds ,$$

i.e.

(1.12) $$|v(t)| \leq \|v\|_{W(0,T)} .$$

Let $t \in [0,\frac{T}{3}[$. Then $\alpha(t) = 1$ and since α is very smooth, (1.12) yields

(1.13) $$|u(t)| \leq C \|u\|_{W(0,T)} \quad \forall\, t \in [0, \frac{T}{3}[.$$

Of course, by applying the same argument to w, we derive (1.13) on the inte $]\frac{2T}{3},T]$. And finally, by repeating this process with another appropriate pai of functions α and β, we obtain (1.13) on the whole interval $[0,T]$.

As far as (1.10) is concerned, it is obviously satisfied for all element u and v of $\mathcal{C}^\infty([0,T];V)$. By virtue of Lemma 1.2 and (1.9), this carr over to all u and v of $W(0,T)$. ■

So far, we have defined spaces which involve integral derivatives of ve valued functions, but it is sometimes useful to work with spaces of fraction derivatives. These are defined by means of a Fourier transform like in section 2.1, chapter IV . More precisely if $t \longmapsto u(t)$ is a vector-valued function on $[0,T]$ and if :

$$\tilde{u}(t) = u(t) \quad \text{on} \quad [0,T], \quad \tilde{u}(t) = 0 \text{ elsewhere},$$

then, the Fourier transform $\tau \longmapsto \hat{u}(\tau)$ is defined on $\mathbb{R}$ by

$$\hat{u}(\tau) = \int_{\mathbb{R}} e^{-2i\pi t\tau}\, \tilde{u}(\tau)d\tau .$$

For $\gamma \in \mathbb{R}^+$, we define the space $\mathcal{H}^\gamma(0,T ; V,H)$ by :

$$\mathcal{H}^\gamma(0,T ; V,H) = \{u \in L^2(0,T ; V) \;;\; \tau \longmapsto |\tau|^\gamma \hat{u}(\tau) \in L^2(\mathbb{R} ; H)\} \tag{1.14}$$

normed by

$$\|u\|_{\mathcal{H}^\gamma} = \{\int_0^T \|u(t)\|^2 dt + \int_{\mathbb{R}} |\tau|^{2\gamma} |\hat{u}(\tau)|^2 d\tau\}^{1/2} .$$

The following lemma will be useful later on .

LEMMA 1.3.

If the imbedding of V into H is compact, then the canonical imbedding of $\mathcal{H}^\gamma(0,T ; V,H)$ into $L^2(0,T ; H)$ is also compact.

PROOF.

Let (u_m) be a bounded sequence in $\mathcal{H}^\gamma(0,T ; V,H)$; we must extract a subsequence (u_μ) from (u_m) that converges in $L^2(0,T ; H)$.

We know that there exist an element u of $\mathcal{H}^\gamma(0,T ; V,H)$ and a subsequence (u_μ) of (u_m) such that

$$\text{weak} \lim_{\mu \to \infty} u_\mu = u \quad \text{in} \quad \mathcal{H}^\gamma(0,T ; V,H)$$

and we must establish that

$$\lim_{\mu \to \infty} u_\mu = u \quad \text{in} \quad L^2(0,T ; H) .$$

First, there is no loss of generality in assuming that $u = 0$. Next, it is equivalent and more convenient to prove that

$$\lim_{\mu \to \infty} \hat{u}_\mu = 0 \quad \text{in} \quad L^2(\mathbb{R} ; H)$$

where $\hat{u}_\mu$ is the Fourier transform of u_μ as defined above. For this, let us write :

$$\int_{\mathbb{R}} |\hat{u}_\mu(\tau)|^2 d\tau = \int_{|\tau| \leq M} |\hat{u}_\mu(\tau)|^2 d\tau + \int_{|\tau| \geq M} (1+|\tau|^{2\gamma}) |u_\mu(\tau)|^2 \frac{d\tau}{1+|\tau|^{2\gamma}} ,$$

where M is any positive number. Since $\| u_m \|_{\mathcal{H}^\gamma}$ is bounded, we get the following bound :

$$\int_{\mathbb{R}} |\hat{u}_\mu(\tau)|^2 d\tau \leq \int_{|\tau| \leq M} |\hat{u}_\mu(\tau)|^2 d\tau + \frac{C_1}{1+M^{2\gamma}} .$$

Thus, since $\lim\limits_{M \to \infty} \frac{1}{1+M^{2\gamma}} = 0$, we are led to show that

$$\text{(1.15)} \qquad \lim_{\mu \to \infty} \int_{|\tau| \leq M} |\hat{u}_\mu(\tau)|^2 d\tau = 0 \qquad \forall\, M < \infty .$$

Now, let w belong to V and for each real τ consider

$$\lim_{\mu \to \infty} (\hat{u}_\mu(\tau), w) = \lim_{\mu \to \infty} \int_0^T (u_\mu(t), e^{-2i\pi t\tau} w) dt = 0$$

since $e^{-2i\pi t\tau} w \in L^2(0,T ; V)$ and weak $\lim\limits_{\mu \to \infty} u_\mu = 0$ in $L^2(0,T ; H)$.

But, for each τ , $\hat{u}_\mu(\tau) \in V$ and $\| \hat{u}_\mu(\tau) \|_V \leq T^{1/2} \| u_\mu \|_{L^2(0,T ; V)} \leq C_2$.

Therefore, for each τ , weak $\lim\limits_{\mu \to \infty} \hat{u}_\mu(\tau) = 0$ in V . As the imbedding of V into H is compact, this implies that

$$\lim_{\mu \to \infty} \hat{u}_\mu(\tau) = 0 \text{ in } H \quad \forall\, \tau .$$

Therefore, (1.15) holds for all finite M . ∎

1.2. Formulation of the Navier-Stokes problem.

Let Ω be a bounded domain of $\mathbb{R}^n$ (in two or three dimensions) with a Lipschitz continuous boundary Γ .

We have seen in § 5 , Chapter I that the Navier-Stokes equations describing the motion of a viscous and incompressible fluid confined in Ω are :

$$(1.16)\quad \begin{cases} \left.\begin{aligned} &\frac{\partial \vec{u}}{\partial t} - \nu\Delta\vec{u} + \sum_{j=1}^{n} u_j \frac{\partial \vec{u}}{\partial x_j} - \overrightarrow{\text{grad}}\, p = \vec{f} \\ &\text{div}\, \vec{u} = 0 \end{aligned}\right\} \text{ for } (x,t) \in \Omega \times R^+ , \\ \vec{u} = \vec{0} \text{ for } (x,t) \in \Gamma \times R^+ , \\ \vec{u}(0) = \vec{u}_o \text{ for } x \in \Omega , \end{cases}$$

where $\vec{f}$ and $\vec{u}_o$ are two prescribed functions.

We propose to derive an appropriate statement of this problem with the help of the material developed in Section 1.1.

Like in the stationary Stokes and Navier-Stokes cases, we introduce the now familiar spaces :

$$\mathcal{V} = \{\vec{v} \in (\mathcal{D}(\Omega))^n ~;~ \text{div}\, \vec{v} = 0\} ,$$

$$V = \{\vec{v} \in (H_o^1(\Omega))^n ~;~ \text{div}\, \vec{v} = 0\} ,$$

equipped with the norm of $(H_o^1(\Omega))^n$ and

$$H = \{\vec{v} \in (L^2(\Omega))^n ~;~ \text{div}\, \vec{v} = 0 ~,~ \gamma_\nu \vec{v} = 0\} ,$$

equipped with the norm of $(L^2(\Omega))^n$. Recall that $\mathcal{V}$ is dense in H and V, and that the spaces H and V are examples of the abstract spaces used in the preceding section.

For $\vec{u}$, $\vec{v}$ and $\vec{w}$ in $(H^1(\Omega))^n$ we define, as usual, the following forms :

$$a_o(\vec{u},\vec{v}) = (\overrightarrow{\text{grad}}\, \vec{u}, \overrightarrow{\text{grad}}\, \vec{v}) ,$$

$$a_1(\vec{w} ~;~ \vec{u},\vec{v}) = \sum_{i,j=1}^{n} \int_\Omega w_j \frac{\partial u_i}{\partial x_j} v_i \, dx ,$$

$$a(\vec{w} ~;~ \vec{u},\vec{v}) = a_o(\vec{u},\vec{v}) + a_1(\vec{w} ~;~ \vec{u},\vec{v}) .$$

As usual, we denote the scalar product of $L^2(\Omega)$ and $(L^2(\Omega))^n$ (and hence of H) by the same symbol : $(.,.)$.

Now, let us examine the above forms when their arguments also depend upon time : say $\vec{w}$ and $\vec{u} \in L^2(0,T ~;~ V)$. As far as a_o is concerned, consider the function $t \longmapsto A_o\vec{u}(t)$ defined a.e. on $[0,T]$ by :

$$(1.17)\qquad A_o\vec{u}(t) \in V' ~,~ \langle A_o\vec{u}(t),\vec{v}\rangle = a_o(\vec{u}(t),\vec{v}) \qquad \forall\, \vec{v} \in V .$$

It can be readily checked that $t \longmapsto A_o\vec{u}(t) \in L^2(0,T\ ;\ V')$ and that $A_o \in \mathcal{L}(L^2(0,T\ ;\ V)\ ;\ L^2(0,T\ ;\ V'))$.

Next, consider the mapping $t \longmapsto A_1(\vec{w}(t),\vec{u}(t))$ defined a.e. on $[0,T]$ by :

$$(1.18)\quad \begin{cases} A_1(\vec{w}(t),\vec{u}(t)) \in V', \\ < A_1(\vec{w}(t),\vec{u}(t)),\vec{v} > = a_1(\vec{w}(t)\ ;\ \vec{u}(t),\vec{v}) \quad \forall\ \vec{v} \in V. \end{cases}$$

Clearly, the function $t \longmapsto A_1(\vec{w}(t),\vec{u}(t))$ is measurable on $(0,T)$, but in order to derive appropriate estimates for $\|A_1(\vec{w}(t),\vec{u}(t))\|_*$, we require first the following preliminary result.

<u>LEMMA 1.4.</u>

<u>When</u> $n = 2$, <u>all elements</u> φ <u>of</u> $H_o^1(\Omega)$ <u>satisfy</u> :

$$(1.19)\quad \|\varphi\|_{o,4,\Omega} \leq 2^{1/4}|\varphi|_{1,\Omega}^{1/2}\|\varphi\|_{o,\Omega}^{1/2}.$$

<u>PROOF</u>.

It suffices to prove (1.19) in $\mathcal{D}(\Omega)$. Let $\tilde{\varphi}$ extend φ by zero outside Ω; then we can write :

$$\tilde{\varphi}^2(x_1,x_2) = 2\int_{-\infty}^{x_1} \tilde{\varphi}(\xi_1,x_2)\,\frac{\partial\tilde{\varphi}}{\partial\xi_1}(\xi_1,x_2)d\xi_1 \leq \varphi_1(x_2),$$

where

$$\varphi_1(x_2) = 2\int_{-\infty}^{\infty} |\tilde{\varphi}(\xi_1,x_2)|\left|\frac{\partial\tilde{\varphi}}{\partial\xi_1}(\xi_1,x_2)\right|d\xi_1.$$

Similarly,

$$\tilde{\varphi}^2(x_1,x_2) \leq \varphi_2(x_1),$$

where

$$\varphi_2(x_1) = 2\int_{-\infty}^{\infty} |\tilde{\varphi}(x_1,\xi_2)|\left|\frac{\partial\tilde{\varphi}}{\partial\xi_2}(x_1,\xi_2)\right|d\xi_2.$$

Hence

$$\int_{\mathbb{R}^2}\tilde{\varphi}^4(x)dx \leq \int_{\mathbb{R}}\varphi_1(x_2)dx_2\int_{\mathbb{R}}\varphi_2(x_1)dx_1.$$

Therefore

$$\|\varphi\|_{o,4,\Omega}^4 \leq 4\,\|\varphi\|_{o,\Omega}^2\,\left\|\frac{\partial\varphi}{\partial x_1}\right\|_{o,\Omega}\left\|\frac{\partial\varphi}{\partial x_2}\right\|_{o,\Omega} \leq 2\,\|\varphi\|_{o,\Omega}^2\,|\varphi|_{1,\Omega}^2. \quad \blacksquare$$

LEMMA 1.5.

When $\vec{w}$ and $\vec{u}$ belong both to $L^2(0,T\,;V) \cap L^\infty(0,T\,;H)$ then

$$(1.20) \qquad A_1(\vec{w},\vec{u}) \in \begin{cases} L^2(0,T\,;V') & \text{if } n = 2 \\ L^{4/3}(0,T\,;V') & \text{if } n = 3\,. \end{cases}$$

PROOF.

As $a_1(\vec{w}\,;\vec{u},\vec{v}) = - a_1(\vec{w}\,;\vec{v},\vec{u})$, we have the upper bound :

$$|a_1(\vec{w}\,;\vec{u},\vec{v})| \leq C_1 \|\vec{w}\|_{o,4,\Omega} \|\vec{u}\|_{o,4,\Omega} |\vec{v}|_{1,\Omega}\,.$$

Therefore

$$(1.21) \qquad \|A_1(\vec{w}(t),\vec{u}(t))\|_\star \leq C_1 \|\vec{w}(t)\|_{o,4,\Omega} \|\vec{u}(t)\|_{o,4,\Omega} \text{ a.e in } (0,T).$$

Now, the argument splits into two cases according that $n = 2$ or 3.

1) When $n = 3$, each function φ of $H_o^1(\Omega)$ satisfies :

$$(1.22) \qquad \|\varphi\|_{o,4,\Omega} \leq \|\varphi\|^{3/4}_{o,6,\Omega} \|\varphi\|^{1/4}_{o,\Omega} \leq C_2 |\varphi|^{3/4}_{1,\Omega} \|\varphi\|^{1/4}_{o,\Omega}\,,$$

owing to Sobolev's imbedding Theorem. Therefore, (1.21) implies that :

$$\|A_1(\vec{w}(t),\vec{u}(t))\|^{4/3}_\star \leq C_3 |\vec{w}(t)|_{1,\Omega} |\vec{u}(t)|_{1,\Omega} \|\vec{w}(t)\|^{1/3}_{o,\Omega} \|\vec{u}(t)\|^{1/3}_{o,\Omega}\,.$$

As a consequence,

$$\int_0^T \|A_1(\vec{w}(t),\vec{u}(t))\|^{4/3}_\star dt \leq C_3 (\|\vec{w}\|^{1/3}_{L^\infty(0,T\,;H)} \|\vec{u}\|^{1/3}_{L^\infty(0,T\,;H)}) \int_0^T |\vec{w}(t)|_{1,\Omega} |\vec{u}(t)|_{1,\Omega} dt\,.$$

Since $\vec{w}$ and $\vec{u}$ belong to $L^\infty(0,T\,;H) \cap L^2(0,T\,;V)$ this implies that

$$\int_0^T \|A_1(\vec{w}(t),\vec{u}(t))\|^{4/3}_\star\, dt \leq C_4\,.$$

2) When $n = 2$, we make use of Lemma 1.4. According to (1.19) and (1.21), we have :

$$\|A_1(\vec{w}(t),\vec{u}(t))\|^2_\star \leq 2\, |\vec{w}(t)|_{1,\Omega} |\vec{u}(t)|_{1,\Omega} \|\vec{w}(t)\|_{o,\Omega} \|\vec{u}(t)\|_{o,\Omega}\,.$$

Therefore,

$$\int_0^T \|A_1(\vec{w}(t),\vec{u}(t))\|^2_\star dt$$

$$\leq 2 \|\vec{w}\|_{L^\infty(0,T\,;H)} \|\vec{u}\|_{L^\infty(0,T\,;H)} \|\vec{w}\|_{L^2(0,T\,;V)} \|\vec{u}\|_{L^2(0,T\,;V)}\,.$$

This proves (1.20). ■

Remark 1.1.

In the course of this proof we have shown that, if $\vec{u}$ belongs to $L^2(0,T\,;\,V) \cap L^\infty(0,T\,;\,H)$ then

(1.23a) $\quad \|A_1(\vec{u},\vec{u})\|_{L^{4/3}(0,T;V')} \leq C\,\|\vec{u}\|^{1/2}_{L^\infty(0,T\,;\,H)}\,\|\vec{u}\|^{3/2}_{L^2(0,T\,;\,V)}$ when $n = 3$,

(1.23b) $\quad \|A_1(\vec{u},\vec{u})\|_{L^2(0,T;V')} \leq C\,\|\vec{u}\|_{L^\infty(0,T;H)}\,\|\vec{u}\|_{L^2(0,T;V)}$ when $n = 2$. ■

With these spaces and forms, consider the following variational formulation of problem (1.16).

For a given function $\vec{f}$ in $L^2(0,T\,;(H^{-1}(\Omega))^n)$ and a given element $\vec{u}_o$ of H , find $\vec{u}$ in $L^2(0,T\,;\,V) \cap L^\infty(0,T\,;\,H)$ such that

$$
(P)\left\{
\begin{array}{l}
(1.24)\ \dfrac{d}{dt}(\vec{u}(t),\vec{v}) + a(\vec{u}(t)\,;\,\vec{u}(t),\vec{v}) = \langle \vec{f}(t),\vec{v}\rangle \text{ in } \mathcal{D}'(]0,T[),\ \forall\,\vec{v}\in V \\[2ex]
\qquad\qquad \text{with the initial condition } \vec{u}(0) = \vec{u}_o .
\end{array}
\right.
$$

Remark 1.2.

1) It may seem unnatural at the outset to look for a solution in $L^\infty(0,T\,;\,H)$ but it will be proved further on that such a solution does exist.

2) The initial condition makes sense only if the solution $\vec{u}$ is continuous at $t = 0$. In fact, it is shown below that $\vec{u}$ is continuous on $[0,T]$. ■

THEOREM 1.2

Let $\vec{u} \in L^2(0,T\,;\,V) \cap L^\infty(0,T\,;\,H)$ be a solution of (1.24). Then

$$
(1.25) \qquad \frac{d\vec{u}}{dt} \in \left\{
\begin{array}{ll}
L^2(0,T\,;\,V') & \text{when } n = 2 \\
L^{4/3}(0,T\,;\,V') & \text{when } n = 3 .
\end{array}
\right.
$$

PROOF.

With the above notation, we have

$$a(\vec{u}(t)\,;\,\vec{u}(t),\vec{v}) = \langle A_o\vec{u}(t) + A_1(\vec{u}(t),\vec{u}(t)),\vec{v}\rangle \quad \forall\,\vec{v}\in V .$$

Therefore, by Lemma 1.1, each solution $\vec{u}$ of (1.24) satisfies in $\mathcal{D}'(]0,T[)$:

$$< \frac{d\vec{u}}{dt}(t),\vec{v} > = < \vec{f}(t) - A_o\vec{u}(t) - A_1(\vec{u}(t),\vec{u}(t)),\vec{v} > \quad \forall \vec{v} \in V .$$

Now, by hypothesis $\vec{f} \in L^2(0,T ; V')$ and we have mentioned previously that $A_o\vec{u} \in L^2(0,T ; V')$ when $\vec{u} \in L^2(0,T ; V)$. Furthermore, $A_1(\vec{u},\vec{u})$ belongs to $L^2(0,T ; V')$ if $n = 2$ or $L^{4/3}(0,T ; V')$ if $n = 3$. Hence

$$\frac{d\vec{u}}{dt} \in \begin{cases} L^2(0,T ; V') & \text{if } n = 2 \\ L^{4/3}(0,T ; V') & \text{if } n = 3 . \end{cases}$$ ■

According to Theorems 1.2 and 1.1, $\vec{u} \in \mathcal{C}^o([0,T] ; H)$ when $n = 2$. If $n = 3$, by applying Theorem 1.2 and the last part of Lemma 1.1, we only get $\vec{u}$ in $\mathcal{C}^o([0,T] ; V')$. In both cases, it is perfectly allowable to prescribe the value of $\vec{u}$ at $t = 0$. Moreover, it stems from the above proof that problem (P) can be equivalently stated as follows :

(P) Given $\vec{f}$ in $L^2(0,T ; (H^{-1}(\Omega))^n)$ and $\vec{u}_o$ in H, find

$$(1.26) \quad \vec{u} \in L^2(0,T ; V) \cap L^\infty(0,T ; H) \text{ with } \frac{d\vec{u}}{dt} \in \begin{cases} L^2(0,T ; V') & \text{if } n = 2 \\ L^{4/3}(0,T ; V') & \text{if } n = 3 \end{cases}$$

such that

$$(1.27) \quad \frac{d\vec{u}}{dt} + A_o\vec{u} + A_1(\vec{u},\vec{u}) = \vec{f} ,$$

$$(1.28) \quad \vec{u}(0) = \vec{u}_o .$$

Remark 1.3.

The proof of Theorem 1.2 shows that (1.24) holds in $L^2(0,T)$ when $n = 2$ or $L^{4/3}(0,T)$ when $n = 3$. Likewise, (1.27) holds in $L^2(0,T ; V')$ if $n = 2$ and $L^{4/3}(0,T ; V')$ if $n = 3$. ■

It remains to verify that this problem is indeed the same as our original problem (1.16). Since (1.27) is essentially obtained by multiplying (1.16) with a divergence-free test function, it suffices to recover the pressure lost in the process. For this, consider the following problem :

$$
(Q)\left\{\begin{array}{ll}
\multicolumn{2}{l}{\text{For } \vec{f} \text{ and } \vec{u}_o \text{ given as above, find a pair } (\vec{u},p) \text{ such that}} \\
(1.29) & \vec{u} \text{ satisfies } (1.26) \quad , \quad p \in \mathcal{D}'(\Omega \times]0,T[) , \\
(1.30) & \dfrac{\partial \vec{u}}{\partial t} - \nu\Delta\vec{u} + \sum_{j=1}^{n} u_j \dfrac{\partial \vec{u}}{\partial x_j} + \overrightarrow{\text{grad}}\, p = \vec{f} \text{ in } \mathcal{D}'(\Omega \times]0,T[) , \\
(1.28) & \vec{u}(0) = \vec{u}_o .
\end{array}\right.
$$

THEOREM 1.3.

Problems (P) and (Q) are equivalent.

PROOF.

Clearly, if $(\vec{u},p)$ is a solution of (Q), then $\vec{u}$ satisfies (P).

Conversely, let $\vec{u} \in L^2(0,T ; V) \cap L^\infty(0,T ; H)$ be a solution of (P) and consider the mapping defined on $(H_o^1(\Omega))^n$ by :

$$L(\vec{v},t) : \vec{v} \longmapsto \int_0^t \{< \vec{f}(s),\vec{v} > - a(\vec{u}(s) \; ; \vec{u}(s),\vec{v})\}ds - (\vec{u}(t),\vec{v}) + (\vec{u}_o,\vec{v}) .$$

For each t, L is a linear functional on $(H_o^1(\Omega))^n$ that vanishes on V .

Hence, according to Theorem 3.6 , Chapter I , for each t there exists exactly one function $P(t) \in L_o^2(\Omega)$, such that

$$L(\vec{v},t) = - < \overrightarrow{\text{grad}}\, P(t),\vec{v} > \qquad \forall\, \vec{v} \in (H_o^1(\Omega))^n .$$

In other words,

$$(1.31) \quad (P(t),\text{div}\, \vec{v}) = \int_0^t \{< \vec{f}(s),\vec{v} > - a(\vec{u}(s);\vec{u}(s),\vec{v})\}ds - (\vec{u}(t),\vec{v}) + (\vec{u}_o,\vec{v}) \qquad \forall\, \vec{v} \in (H_o^1(\Omega))^n.$$

By using Lemma 1.5, it can be checked that $P \in \mathcal{C}^o([0,T];L_o^2(\Omega))$.

Next, by differentiating (1.31), we get :

$$(1.32) \quad < \frac{dP}{dt}(t),\text{div}\, \vec{v} > = < \vec{f}(t),\vec{v} > - a(\vec{u}(t) \; ; \vec{u}(t),\vec{v}) - < \frac{d\vec{u}(t)}{dt},\vec{v} > \qquad \forall\, v \in (H_o^1(\Omega))^n .$$

Thus, if we set $p = \dfrac{dP}{dt}$ in $\mathcal{D}'(\Omega \times]0,T[)$, we find (1.30) . ∎

Remark 1.4.

From (1.32), we derive immediately :

$$\frac{d}{dt} < \vec{u}(t) - \overrightarrow{\text{grad}}\, P(t),\vec{v} > + \, a(\vec{u}(t) \; ; \; \vec{u}(t),\vec{v}) = < \vec{f}(t),\vec{v} > \quad \forall \, \vec{v} \in (H_o^1(\Omega))^n \, .$$

It can be checked that this implies :

$$(1.33) \qquad \frac{d}{dt}(\vec{u} + \overrightarrow{\text{grad}}\, P) \in \begin{cases} L^2(0,T \; ; \; (H^{-1}(\Omega))^n) \\ L^{4/3}(0,T \; ; \; (H^{-1}(\Omega))^n) \end{cases} ;$$

however this furnishes no precision either about $\frac{d\vec{u}}{dt}$ or $\frac{d}{dt}(\overrightarrow{\text{grad}}\, P)$ alone . ■

1.3. Existence and uniqueness of the solution.

In order to prove that problem (P) has a solution, we propose to construct first a sequence (P_m) of semi-discrete problems, each of which has a unique solution. Then, by means of a priori estimates, we show that some subsequence of these solutions converges toward a function that satisfies (P). More precisely, we consider a basis $(\vec{w}_m)_{m \geq 1}$ of V (cf. Theorem 1.2, Chapter IV) and we denote by V_m the space spanned by the set $\{\vec{w}_1 \, ,\ldots, \, \vec{w}_m\}$. Then we replace problem (P) by the following problem in $V_m \times [\,0,T\,]$:

Find a function $\vec{u}_m(t)$ of the form $\vec{u}_m(t) = \sum_{j=1}^{m} (\, g_{jm}(t)\vec{w}_j$

satisfying the initial value system of ordinary differential equations :

$$(P_m) \begin{cases} (1.34) \quad \frac{d}{dt}(\vec{u}_m(t),\vec{w}_i) + a(\vec{u}_m(t) \; ; \; \vec{u}_m(t),\vec{w}_i) = < \vec{f}(t),\vec{w}_i > \quad \text{for} \quad 1 \leq i \leq m \, , \\ \vec{u}_m(0) = \vec{u}_{om} \in V_m \, . \end{cases}$$

The starting value $\vec{u}_{om}$ is chosen so that $\lim_{m \to \infty} \vec{u}_{om} = \vec{u}_o$ in H .

For example, $\vec{u}_{om}$ can be the projection of $\vec{u}_o$ on V_m for the norm of H .

LEMMA 1.6.

Problem (P_m) has a unique solution $\vec{u}_m$ in $L^\infty(0,T\,;H) \cap L^2(0,T\,;V)$.

Moreover,

$$\|\vec{u}_m\|_{L^\infty(0,T\,;\,H)} + \|\vec{u}_m\|_{L^2(0,T\,;\,V)} \leq C,$$

where C is a constant independent of m.

PROOF.

Let us write problem (P_m) in terms of its unknowns $g_{jm}(t)$:

$$(1.35)\qquad \sum_{j=1}^{m} (\vec{w}_j,\vec{w}_i)\frac{d}{dt}g_{jm}(t) + \sum_{j=1}^{m}\sum_{k=1}^{m} a(\vec{w}_j\,;\vec{w}_k,\vec{w}_i)g_{jm}(t)g_{km}(t) = \langle\vec{f}(t),\vec{w}_i\rangle$$

$$\text{for}\quad 1 \leq i \leq m$$

with the starting value

$$(1.36)\qquad g_{jm}(0) = g^o_{jm} \qquad \text{for}\quad 1 \leq j \leq m,$$

where g^o_{jm} are the coefficients of $\vec{u}_{om}$.

The $m \times m$ matrix $[(\vec{w}_j,\vec{w}_i)]$ for $1 \leq i,j \leq m$ is nonsingular since $\vec{w}_1,\ldots,\vec{w}_m$ are linearly independent. Therefore (1.35) is a system of the form

$$(1.35')\qquad \frac{d}{dt}g_{im}(t) = \varphi_i(t\,;g_{1m}(t),\ldots,g_{mm}(t)) \quad \text{for}\quad 1 \leq i \leq m,$$

with (1.36) unchanged. Now, according to Carathéodory's theorem (cf. [16]), this system of ordinary differential equations has a local maximal solution $\vec{u}_m(t)$ in an interval $[0,t_m[$ for some $t_m \leq T$. If $t_m < T$ then necessarily,

$$\lim_{t\to t_m-} |\vec{u}_m(t)| = \infty.$$

Therefore, if we show that $|\vec{u}_m(t)|$ is bounded independently of m and t, then this will prove that $t_m = T \;\forall\, m$ and that $\vec{u}_m(t)$ is in fact a global solution. For this, let $t \in [0,t_m[$, multiply (1.34) by $g_{im}(t)$ and sum over i from 1 to m. In view of (2.7) Chapter IV, we get :

$$\left(\frac{d}{dt}\vec{u}_m(t),\vec{u}_m(t)\right) + \nu\,\|\vec{u}_m(t)\|^2 = \langle \vec{f}(t),\vec{u}_m(t)\rangle .$$

Next, let us integrate both sides of this equation and apply Green's formula (1.10):

$$(1.37)\qquad \frac{1}{2}\,|\vec{u}_m(t)|^2 - \frac{1}{2}|\vec{u}_m(0)|^2 + \nu\int_0^t \|\vec{u}_m(s)\|^2 ds = \int_0^t \langle \vec{f}(s),\vec{u}_m(s)\rangle\, ds .$$

But for all $\varepsilon > 0$, the right-hand side is bounded as follows :

$$(1.38)\qquad \left|\int_0^t \langle \vec{f}(s),\vec{u}_m(s)\rangle\, ds\right| \leq \frac{1}{2}\left\{\varepsilon\int_0^t \|\vec{u}_m(s)\|^2 ds + \frac{1}{\varepsilon}\int_0^t \|\vec{f}(s)\|_\star^2\, ds\right\} .$$

Therefore, by choosing $\varepsilon = 2\nu$, (1.37),(1.38) yield the upper bound :

$$|\vec{u}_m(t)|^2 \leq |\vec{u}_{om}|^2 + \frac{1}{2\nu}\int_0^t \|\vec{f}(s)\|_\star^2\, ds ,$$

which, in turn, can be bounded independently of m and t . Therefore $\vec{u}_m$ is bounded in $L^\infty(0,T\,;\,H)$.

In addition, by substituting (1.38) with $\varepsilon = \nu$ in (1.37), we obtain (with $t = T$) :

$$|\vec{u}_m(T)|^2 + \nu\int_0^T \|\vec{u}_m(t)\|^2 dt \leq |\vec{u}_{om}|^2 + \frac{1}{\nu}\int_0^T \|\vec{f}(t)\|_\star^2\, dt .$$

Hence $\vec{u}_m$ is bounded in $L^2(0,T\,;\,V)$.

Finally, it is easy to prove that $\vec{u}_m$ is the only solution of (P_m) . ∎

The following lemma gives useful information about some fractional derivatives of $\vec{u}_m$ with respect to time.

LEMMA 1.7.

The sequence $(\vec{u}_m)$ is bounded in $\mathcal{H}^\gamma(0,T\,;\,V,H)$ for $0 < \gamma < \frac{1}{4}$.

PROOF.

According to the definition of $\mathcal{H}^\gamma(0,T\,;\,V,H)$, we first extend $\vec{u}_m(t)$ by zero outside $[0,T]$ and then we take the Fourier transform of the extended function $\tilde{\vec{u}}_m(t)$. At the first stage, the system (1.34) is extended to the whole line as follows :

$$(1.39)\quad \frac{d}{dt}(\tilde{\vec{u}}_m(t),\vec{w}_i) + a(\tilde{\vec{u}}_m(t)\ ;\ \tilde{\vec{u}}_m(t),\vec{w}_i) = \langle \tilde{\vec{f}}(t),\vec{w}_i \rangle$$
$$+ (\vec{u}_{om},\vec{w}_i)\delta_o - (\vec{u}_m(T)\ ,\ \vec{w}_i)\delta_T \quad \text{for } 1 \leq i \leq m\ ,$$

where δ_o and δ_T denote respectively the Dirac distribution at $t = 0$ and at $t = T$ and $\tilde{\vec{f}}$ is the extension of $\vec{f}$ by zero.

For the second step, it is shorter to write :

$$\vec{g}(\vec{u}) = A_1(\vec{u},\vec{u})\ .$$

Also, we denote by $\hat{\vec{u}}_m(\tau)$, $\hat{\vec{f}}(\tau)$ and $\hat{\vec{g}}_m(\tau)$ respectively the Fourier transform of $\tilde{\vec{u}}_m(t)$, $\tilde{\vec{f}}(t)$ and $\vec{g}(\tilde{\vec{u}}_m(t))$. Then, we take the Fourier transform of both sides of (1.39) ; we get :

$$2\ i\pi\tau(\hat{\vec{u}}_m(\tau),\vec{w}_j) + a_o(\hat{\vec{u}}_m(\tau),\vec{w}_j) + \langle \hat{\vec{g}}_m(\tau),\vec{w}_j \rangle = \langle \hat{\vec{f}}(\tau),\vec{w}_j \rangle$$
$$+ (\vec{u}_{om},\vec{w}_j) - (\vec{u}_m(T),\vec{w}_j)e^{-2i\pi\tau T} \quad \text{for } 1 \leq j \leq m.$$

This implies that :

$$2i\pi\tau|\hat{\vec{u}}_m(\tau)|^2 + \nu\,\|\hat{\vec{u}}_m(\tau)\|^2 + \langle \hat{\vec{g}}_m(\tau),\hat{\vec{u}}_m(\tau) \rangle = \langle \hat{\vec{f}}(\tau),\hat{\vec{u}}_m(\tau) \rangle$$
$$+ (\vec{u}_{om},\hat{\vec{u}}_m(\tau)) - (\vec{u}_m(T),\hat{\vec{u}}_m(\tau))e^{-2i\pi\tau T}\ .$$

The imaginary part of this equality yields the following upper bound :

$$2\pi|\tau|\,|\hat{\vec{u}}_m(\tau)|^2 \leq (\|\hat{\vec{g}}_m(\tau)\|_\star + \|\hat{\vec{f}}(\tau)\|_\star)\|\hat{\vec{u}}_m(\tau)\|$$
$$+ (|\vec{u}_{om}| + |\vec{u}_m(T)|)|\hat{\vec{u}}_m(\tau)|\ .$$

But

$$\|\hat{\vec{f}}(\tau)\|_\star \leq \int_{-\infty}^{+\infty} \|\tilde{\vec{f}}(t)\|_\star dt \leq C_1\ .$$

Moreover, according to (1.21) ,

$$\|\hat{\vec{g}}_m(\tau)\|_\star \leq \int_{-\infty}^{+\infty} \|A_1(\tilde{\vec{u}}_m(t),\tilde{\vec{u}}_m(t))\|_\star\ dt \leq C_2 \int_0^T \|\vec{u}_m(t)\|^2\ dt\ .$$

Therefore,

$$\|\hat{\vec{g}}_m(\tau)\|_\star \leq C_3\ .$$

Hence, the following bound holds for all $\tau \in \mathbb{R}$:

(1.40) $$|\tau|\hat{\vec{u}}_m(\tau)|^2 \leqslant C_4(\|\hat{\vec{u}}_m(\tau)\| + |\hat{\vec{u}}_m(\tau)|) .$$

Now, let us divide both sides of (1.40) by $(1+|\tau|^\sigma)$ for some σ with $\frac{1}{2} < \sigma < 1$ and let us integrate over $\mathbb{R}$. We have :

$$\int_{-\infty}^{+\infty} \frac{\|\hat{\vec{u}}_m(\tau)\|}{1+|\tau|^\sigma} d\tau \leqslant \|\vec{u}_m\|_{L^2(0,T\,;\,V)} \left[\int_{-\infty}^{+\infty} \frac{d\tau}{(1+|\tau|^\sigma)^2}\right]^{1/2}$$

$$\leqslant C_5 \|\vec{u}_m\|_{L^2(0,T\,;\,V)} \quad \text{since } \sigma > \frac{1}{2} .$$

Similarly,

$$\int_{-\infty}^{+\infty} \frac{|\hat{\vec{u}}_m(\tau)|}{1+|\tau|^\sigma} d\tau \leqslant C_5 \|\vec{u}_m\|_{L^2(0,T\,;\,H)} .$$

Therefore, by Lemma 1.6 , we get

$$\int_{-\infty}^{+\infty} \frac{|\tau|\hat{\vec{u}}_m(\tau)|^2}{1+|\tau|^\sigma} d\tau \leqslant C_6 \quad \text{for } 1 > \sigma > \frac{1}{2} , \ \forall m .$$

Together with Lemma 1.6, this implies that $\displaystyle\int_{-\infty}^{+\infty} |\tau|^{2\gamma}|\hat{\vec{u}}_m(\tau)|^2 \, d\tau \leqslant C_7$,

for some γ such that $0 < \gamma < \frac{1}{4}$ (for instance , $\gamma = (1-\sigma)/2$) ; this proves the lemma . ■

With the a priori estimates of Lemmas 1.6 and 1.7 , we can prove the following existence result.

<u>THEOREM 1.4.</u>

<u>Problem (P) has at least one solution</u> $\vec{u}$ <u>in</u> $L^2(0,T\,;\,V) \cap L^\infty(0,T\,;\,H)$.

<u>PROOF</u>.

According to Lemmas 1.6 and 1.7 , there exists a subsequence $(\vec{u}_\mu)$ of $(\vec{u}_m)$ such that

(1.41) $$\text{weak} \lim_{\mu \to \infty} \vec{u}_\mu = \vec{u} \quad \text{in } L^2(0,T\,;\,V) ,$$

(1.42) $$\text{weak} \star \lim_{\mu \to \infty} \vec{u}_\mu = \vec{u} \quad \text{in } L^\infty(0,T\,;\,H) ,$$

$$\text{weak} \lim_{\mu \to \infty} \vec{u}_\mu = \vec{u} \quad \text{in } \mathcal{H}^\gamma(0,T\,;\,V,H).$$

Therefore, by Lemma 1.3 , it follows that

(1.43) $$\text{strong} \lim_{\mu \to \infty} \vec{u}_\mu = \vec{u} \text{ in } L^2(0,T ; H) \subset (L^2(\Omega \times]0,T[))^n .$$

These convergences will enable us to pass to the limit in problem (P_μ).

Without loss of generality, we can assume that the basis $(\vec{w}_i)_{i \geq 1} \subset \mathcal{V}$.

Then, take a function ψ in $\mathcal{C}^1([0,T])$ with $\psi(T) = 0$, multiply (1.34) with $\psi(t)$, integrate over $[0,T]$ and use Green's formula (1.10) ; this gives :

(1.44) $$-\int_0^T (\vec{u}_m(t),\vec{w}_i)\psi'(t)dt + \int_0^T a(\vec{u}_m(t) \; ; \vec{u}_m(t),\vec{w}_i)\psi(t)dt = \int_0^T < \vec{f}(t),\vec{w}_i > \psi(t) \, dt + (\vec{u}_{om},\vec{w}_i)\psi(0) , \quad 1 \leq i \leq m .$$

Now, let us fix an arbitrary integer μ_o and let $\vec{v} \in V_{\mu_o}$. Then (1.44) implies :

(1.45) $$-\int_0^T (\vec{u}_\mu(t),\vec{v})\psi'(t)dt + \int_0^T a(\vec{u}_\mu(t) \; ; \vec{u}_\mu(t),\vec{v})\psi(t)dt = \int_0^T < \vec{f}(t),\vec{v} > \psi(t)dt - (\vec{u}_{o\mu},\vec{v})\psi(0) \quad \forall \mu > \mu_o .$$

By virtue of (1.41) , the following limits hold :

$$\lim_{\mu \to \infty} \int_0^T (\vec{u}_\mu(t),\vec{v})\psi'(t)dt = \int_0^T (\vec{u}(t),\vec{v})\psi'(t)dt$$

and

$$\lim_{\mu \to \infty} \int_0^T a_o(\vec{u}_\mu(t),\vec{v})\psi(t)dt = \int_0^T a_o(\vec{u}(t),\vec{v})\psi(t)dt .$$

In addition, since

$$\int_0^T a_1(\vec{u}_\mu(t) \; ; \vec{u}_\mu(t),\vec{v})\psi(t)dt = -\int_0^T a_1(\vec{u}_\mu(t) \; ; \vec{v},\vec{u}_\mu(t))\psi(t)dt = -\int_0^T \left[\sum_{i,j=1}^n \int_\Omega (u_\mu)_j \frac{\partial v_i}{\partial x_j}(u_\mu)_i \; \psi(t) \right] dxdt ,$$

where $\frac{\partial v_i}{\partial x_j} \in \mathcal{D}(\Omega)$, and since the product $(u_\mu)_j(u_\mu)_i$ converges strongly toward $u_j u_i$ in $L^1(\Omega \times]0,T[)$ (owing to (1.43)), it follows that

$$\lim_{\mu \to \infty} \int_0^T a_1(\vec{u}_\mu(t) ; \vec{u}_\mu(t),\vec{v})\psi(t)\,dt = \int_0^T a_1(\vec{u}(t);\vec{u}(t),\vec{v})\psi(t)dt .$$

Finally, by hypothesis, we have : $\lim_{\mu \to \infty} \vec{u}_{o\mu} = \vec{u}_o$ in H.

Hence , as μ tends to infinity , (1.45) becomes :

$$- \int_0^T (\vec{u}(t),\vec{v})\psi'(t)dt + \int_0^T a(\vec{u}(t) ; \vec{u}(t),\vec{v})\psi(t)dt$$

$$(1.46) \qquad = \int_0^T < \vec{f}(t),\vec{v} > \psi(t)dt - (\vec{u}_o,\vec{v})\psi(0) \quad \forall \vec{v} \in \mathbf{V}_{\mu_o} ,$$

$$\forall \psi \in \mathcal{C}^1([0,T]) \text{ with } \psi(T) = 0 .$$

But μ_o is arbitrary and $\bigcup_{m \geq 1} \mathbf{V}_m$ is dense in $\mathbf{V}$. Therefore (1.46) is also valid for all $\vec{v} \in \mathbf{V}$. Furthermore, by restricting ψ to $\mathcal{D}(]0,T[)$, it gives :

$$\frac{d}{dt}(\vec{u}(t),\vec{v}) + a(\vec{u}(t) ; \vec{u}(t),\vec{v}) = < \vec{f}(t),\vec{v} > \quad \forall \vec{v} \in \mathbf{V} , \text{ in } \mathcal{D}'(]0,T[) ,$$

which is precisely (1.24). In fact, as mentioned in Remark 1.3, this equality holds in $L^2(0,T)$ when $n = 2$ or $L^{4/3}(0,T)$ when $n = 3$.

It remains to prove that $\vec{u}(0) = \vec{u}_o$. For this, we multiply (1.24) by a function ψ like in (1.46), integrate over $[0,T]$ and use Green's formula. Comparing with (1.46) we obtain :

$$(\vec{u}(0),\vec{v}) = (\vec{u}_o,\vec{v}) \qquad \forall \vec{v} \in \mathbf{V} .$$

Hence $\qquad \vec{u}(0) = \vec{u}_o$ in $\mathbf{V}'$ – and also in H, since $\vec{u}_o \in H$.

Therefore $\vec{u}$ is a solution of the Navier-Stokes problem (P). ■

As far as the uniqueness of the solution is concerned, let us first establish the following result :

<u>LEMMA 1.8.</u> (Gronwall)

<u>Let</u> m <u>be an integrable and a.e. positive function on</u> $(0,T)$; <u>let</u> $C \geq 0$ <u>be a constant, and let</u> $\varphi \in \mathcal{C}^o([0,T])$ <u>satisfy the inequalities</u> :

$$(1.47) \qquad 0 \leq \varphi(t) \leq C + \int_0^t m(s)\varphi(s)ds \qquad \forall t \in [0,T] .$$

Then φ is bounded as follows :

$$(1.48) \qquad \varphi(t) \leq C \exp\left(\int_0^t m(s)\, ds\right) \quad \forall\, t \in [0,T] .$$

PROOF.

Suppose first that $C > 0$. The inequalities (1.47) imply that

$$0 \leq \frac{m(t)\varphi(t)}{C + \int_0^t m(s)\varphi(s)ds} \leq m(t) \qquad \text{on } [0,T] .$$

After integrating both sides over $(0,t)$ for any $t \in [0,T]$, we get :

$$\mathrm{Log}\left(C + \int_0^t m(s)\varphi(s)ds\right) \leq \mathrm{Log}\left(C \exp\left(\int_0^t m(s)ds\right)\right)$$

i.e.

$$C + \int_0^t m(s)\varphi(s)ds \leq C \exp\left[\int_0^t m(s)ds\right]$$ and the result follows from (1.47).

Hence (1.48) holds whenever $C > 0$ and therefore, it is also valid in the limit when $C = 0$ (in which case $\varphi \equiv 0$). ■

THEOREM 1.5.

When $n = 2$, problem (P) has a unique solution $\vec{u} \in L^2(0,T\,;\,V) \cap L^\infty(0,T\,;\,H)$.

PROOF.

Let $\vec{u}_1$ and $\vec{u}_2$ be two solutions of problem (P) and let $\vec{w} = \vec{u}_1 - \vec{u}_2$. According to (1.27) and Remark 1.3 , $\vec{w}$ satisfies the equation :

$$\frac{d\vec{w}}{dt} + A_o\vec{w} + A_1(\vec{u}_1,\vec{u}_1) - A_1(\vec{u}_2,\vec{u}_2) = 0 \quad \text{in} \quad L^2(0,T\,;\,V') ,$$

$$\vec{w}(0) = \vec{0} .$$

Hence, by taking the scalar product of both sides with $\vec{w}$, we get :

$$(1.49) \qquad \left(\frac{d\vec{w}}{dt},\vec{w}\right) + a_o(\vec{w},\vec{w}) + a_1(\vec{u}_1\,;\,\vec{u}_1,\vec{w}) - a_1(\vec{u}_2\,;\,\vec{u}_2,\vec{w}) = 0 \quad \text{a.e. on} \quad (0,T) .$$

But, by virtue of (2.7) Chapter IV ,

$$a_1(\vec{u}_1\,;\,\vec{u}_1,\vec{w}) - a_1(\vec{u}_2\,;\,\vec{u}_2,\vec{w}) = a_1(\vec{w}\,;\,\vec{u}_1,\vec{w}) \quad \text{a.e. on} \quad (0,T) .$$

Therefore,

$$|a_1(\vec{u}_1 \; ; \vec{u}_1,\vec{w}) - a_1(\vec{u}_2 \; ; \vec{u}_2,\vec{w})| \leq C_1 \|\vec{u}_1\| \, \|\vec{w}\|^2_{o,4,\Omega} \quad \text{a.e. on} \quad (0,T).$$

Hence, by applying Lemma 1.4 and the inequality $ab \leq \nu a^2 + \frac{1}{4\nu} b^2$, we get :

$$|a_1(\vec{u}_1 \; ; \vec{u}_1,\vec{w}) - a_1(\vec{u}_2 \; ; \vec{u}_2,\vec{w})| \leq C_2 \|\vec{u}_1\| \, \|\vec{w}\| \, |\vec{w}|$$

$$\leq \nu \|\vec{w}\|^2 + \frac{C_2^2}{4\nu} \|\vec{u}_1\|^2 |\vec{w}|^2 \, .$$

Let us substitute this last inequality in (1.49) :

$$(\frac{d\vec{w}}{dt}, \vec{w}) \leq \frac{C_2^2}{4\nu} \|\vec{u}_1\|^2 |\vec{w}|^2 \quad \text{a.e. on} \quad (0,T) \, .$$

Since $\vec{w} \in W(0,T)$ (because $n = 2$) , we can integrate the above inequality over $(0,t)$ and apply Green's formula (1.10). It yields :

$$\frac{1}{2} |\vec{w}(t)|^2 \leq \frac{C_2^2}{4\nu} \int_0^t \|\vec{u}_1(s)\|^2 |\vec{w}(s)|^2 ds \, .$$

Now, according to Theorem 1.1 , the mapping $t \longmapsto |\vec{w}(t)|^2$ is continuous on $[0,T]$. Therefore, we can apply Lemma 1.8 with $C = 0$: it implies that $|\vec{w}(t)| = 0$ on $[0,T]$. Hence (P) has a unique solution . ■

§ 2 - NUMERICAL SOLUTION BY SEMI-DISCRETIZATION : A ONE-STEP METHOD

The next two paragraphs are devoted to the numerical solution of the transient Navier-Stokes equations. We focus our attention on the discretization with respect to the time variable, since the discretization with respect to the space variable has been thoroughly studied in Chapter IV. In this paragraph, we propose to analyze a very simple one-step method in order to illustrate the type of argument that is often used when dealing with semi-discretization.

Consider again the problem (P) of § 1 :

Find $\vec{u} \in L^2(0,T ; \mathbf{V}) \cap L^\infty(0,T ; H)$ satisfying

$$(2.1) \quad \begin{cases} \dfrac{d}{dt}(\vec{u}(t),\vec{v}) + a(\vec{u}(t) ; \vec{u}(t),\vec{v}) = \langle \vec{f}(t),\vec{v} \rangle & \forall \vec{v} \in \mathbf{V}, \text{ in } \mathcal{D}'(]0,T[), \\ \vec{u}(0) = \vec{u}_o , \end{cases}$$

where $\vec{f}$ is given in $L^2(0,T ; (H^{-1}(\Omega))^n)$ and $\vec{u}_o$ is given in H.

Let us choose a positive integer N, let k denote the corresponding time-step : $k = T/N$ and t_n the subdivisions of $[0,T]$:

$$t_n = nk \quad , \quad 0 \leq n \leq N .$$

Now, suppose that an approximation, $\vec{u}^n \in \mathbf{V}$, of $\vec{u}(t_n)$ is available and consider the following problem :

$$(2.2) \quad \begin{cases} \text{Find } \vec{u}^{n+1} \in \mathbf{V} \text{ such that} \\ \dfrac{1}{k}(\vec{u}^{n+1}-\vec{u}^n,\vec{v}) + a(\vec{u}^n ; \vec{u}^{n+1},\vec{v}) = \langle \vec{f}^{n+1},\vec{v} \rangle \quad \forall \vec{v} \in \mathbf{V} , \end{cases}$$

where

$$(2.3) \quad \vec{f}^{n+1} = \begin{cases} \vec{f}(t_{n+1}) & \text{if } f \in \mathcal{C}^o([0,T] ; (H^{-1}(\Omega))^n) , \\ \dfrac{1}{k}\displaystyle\int_{t_n}^{t_{n+1}} \vec{f}(t)\, dt & \text{if } \vec{f} \in L^2(0,T ; (H^{-1}(\Omega))^n) . \end{cases}$$

Note that (2.2) is a linear (Stokes-like) problem that changes with each value of n. This means that this semi-discretization of problem (2.1) requires the solution of N distinct linear problems. On the other hand, if in (2.2) , the term $a(\vec{u}^n ; \vec{u}^{n+1},\vec{v})$ is replaced by $a(\vec{u}^{n+1} ; \vec{u}^{n+1},\vec{v})$, the problem becomes non linear and its solution is much more complicated.

LEMMA 2.1.

Let $0 \leq n \leq N-1$. For a given $\vec{u}^n \in \mathbf{V}$, the method (2.2), (2.3) defines a unique $\vec{u}^{n+1} \in \mathbf{V}$.

PROOF.

As $\vec{u}^n$ and $\vec{f}^{n+1}$ are given respectively in $\mathbf{V}$ and in $\mathbf{V}'$, it follows that (2.2) can be expressed in the form :

$$(2.4) \qquad (\vec{u}^{n+1},\vec{v}) + ka(\vec{u}^n ; \vec{u}^{n+1},\vec{v}) = \langle \vec{\ell},\vec{v} \rangle \quad \forall \vec{v} \in \mathbf{V},$$

where $\vec{\ell} \in \mathbf{V}'$. Thus, we are asked to solve a linear boundary value problem associated with the bilinear form :

$$\vec{u},\vec{v} \longmapsto (\vec{u},\vec{v}) + ka(\vec{u}^n ; \vec{u},\vec{v}) .$$

This form is continuous in $\mathbf{V} \times \mathbf{V}$ and $\mathbf{V}$-elliptic since

$$|\vec{v}|^2 + ka(\vec{u}^n ; \vec{v},\vec{v}) = |\vec{v}|^2 + \nu k \|\vec{v}\|^2 .$$

Therefore, by Lax-Milgram's theorem (1.6 Chapter I), problem (2.4) has a unique solution $\vec{u}^{n+1}$ in $\mathbf{V}$. ∎

In order to start the sequence, we *assume that the first approximation* $\vec{u}^o$ *is given in* $\mathbf{V}$. Then the sequence $(\vec{u}^n)_{n=1}^N$ is uniquely defined by (2.2) and (2.3).

Now, we turn to the convergence of the sequence $(\vec{u}^n)$ when the time-step k tends to zero. For this, let us introduce the function $\vec{u}_k \in \mathcal{C}^o([0,T] ; \mathbf{V})$ defined by :

$$\vec{u}_k(t) = \vec{u}^n + \frac{t-nk}{k}(\vec{u}^{n+1} - \vec{u}^n) \quad \forall t \in [nk,(n+1)k],\ 0 \leq n \leq N-1 .$$

We propose to establish that $\vec{u}_k$ converges to a solution $\vec{u}$ of (2.1) in much the same way that we proved the convergence of the sequence $(\vec{u}_m)$ in § 1. That is, we first derive useful a priori estimates and then we pass to the limit, thus deriving an alternate proof of the existence Theorem 1.4.

LEMMA 2.2.

The function $\vec{u}_k$ *satisfies the following discrete a priori estimates :*

$$\max_{0 \leq n \leq N} |\vec{u}^n| \leq C_1 \ , \tag{2.5}$$

$$k\Big(\sum_{n=1}^{N} \|\vec{u}^n\|^2\Big) \leq C_2 \ , \tag{2.6}$$

$$\sum_{n=0}^{N-1} |\vec{u}^{n+1} - \vec{u}^n|^2 \leq C_3 \ , \tag{2.7}$$

where all constants are independent of k .

PROOF.

Let us choose $\vec{v} = \vec{u}^{n+1}$ in (2.2) and use the identity :

$$(a-b)\ a = \frac{1}{2}(a^2 - b^2 + (a-b)^2) \ . \tag{2.8}$$

We obtain

$$\frac{1}{2k}\{|\vec{u}^{n+1}|^2 - |\vec{u}^n|^2 + |\vec{u}^{n+1} - \vec{u}^n|^2\} + \nu \|\vec{u}^{n+1}\|^2 = \langle \vec{f}^{n+1}, \vec{u}^{n+1} \rangle \ .$$

Then, by summing from 0 to $m-1$, this gives $\forall\ \varepsilon > 0$:

$$|\vec{u}^m|^2 - |\vec{u}^o|^2 + \sum_{n=0}^{m-1} |\vec{u}^{n+1} - \vec{u}^n|^2 + 2\ \nu k \sum_{n=1}^{m} \|\vec{u}^n\|^2 \leq 2k \sum_{n=1}^{m} \|\vec{f}^n\|_\star \|\vec{u}^n\|$$

$$\leq \varepsilon k \sum_{n=1}^{m} \|\vec{u}^n\|^2 + \frac{1}{\varepsilon} k \sum_{n=1}^{m} \|\vec{f}^n\|_\star^2 \ .$$

In the general case, when $\vec{f}$ is not continuous, formula (2.3) yields :

$$k \sum_{n=1}^{m} \|\vec{f}^n\|_\star^2 = \frac{1}{k} \sum_{n=1}^{m} \Big\| \int_{t_{n-1}}^{t_n} \vec{f}(t)dt \Big\|_\star^2 \leq \int_0^{t_m} \|\vec{f}(t)\|_\star^2 dt \ .$$

Hence, for $1 \leq m \leq N$:

$$|\vec{u}^m|^2 + \sum_{n=0}^{m-1} |\vec{u}^{n+1} - \vec{u}^n|^2 + 2\nu k \sum_{n=1}^{m} \|\vec{u}^n\|^2 \leq |\vec{u}^o|^2 + \varepsilon k \sum_{n=1}^{m} \|\vec{u}^n\|^2 + \frac{1}{\varepsilon} \int_o^{t_m} \|\vec{f}(t)\|_\star^2 dt \ . \tag{2.9}$$

With $\varepsilon = 2\nu$, (2.9) becomes :

$$|\vec{u}^m|^2 + \sum_{n=0}^{m-1} |\vec{u}^{n+1} - \vec{u}^n|^2 \leq |\vec{u}^o|^2 + \frac{1}{2\nu} \int_o^{t_m} \|\vec{f}(t)\|_\star^2 dt \ , \quad 1 \leq m \leq N \ .$$

As the right-hand side is bounded independently of m , this implies that :

$$\max_{0 \leqslant m \leqslant N} |\vec{u}^m| \leqslant C_1 \quad , \quad \sum_{n=0}^{N-1} |\vec{u}^{n+1} - \vec{u}^n|^2 \leqslant C_2 \ .$$

Similarly, with $\varepsilon = \nu$ and $m = N$, (2.9) yields $\nu k \sum_{n=1}^{N} \|\vec{u}^n\|^2 \leqslant C_3$. ■

LEMMA 2.3.

The sequence of functions $(\vec{u}_k)$ is bounded in $\mathcal{H}^{\gamma}(0,T\ ;\ \mathbf{V},H) \cap L^{\infty}(0,T\ ;\ H)$ with $0 < \gamma < \frac{1}{4}$.

PROOF.

It follows immediately from (2.5) (resp.(2.6)) that $(\vec{u}_k)$ is bounded in $L^{\infty}(0,T\ ;\ H)$ (resp. $L^2(0,T\ ;\ \mathbf{V})$). Therefore, it suffices to examine the behavior of some fractional derivative of $\vec{u}_k$. For this, we introduce the step-functions $\vec{g}_k$ and $\vec{f}_k \in L^2(0,T\ ;\ \mathbf{V}')$ defined respectively by :

$$\left.\begin{aligned} &< \vec{g}_k(t),\vec{v} > = a_o(\vec{u}^{n+1},\vec{v}) + a_1(\vec{u}^n\ ;\ \vec{u}^{n+1},\vec{v}) \quad \forall\, \vec{v} \in \mathbf{V} \\ &\vec{f}_k(t) = \vec{f}^{n+1} \ . \end{aligned}\right\} \begin{aligned} &t \in \left]t_n,t_{n+1}\right] , \\ &0 \leqslant n \leqslant N-1 \ . \end{aligned}$$

Next, we extend $\vec{u}_k$, $\vec{g}_k$ and $\vec{f}_k$ by zero outside [0,T]. Then (2.2) has the following equivalent formulation :

$$\frac{d}{dt}(\tilde{\vec{u}}_k(t),v) + < \tilde{\vec{g}}_k(t),\vec{v} > = < \tilde{\vec{f}}_k(t),\vec{v} > + (\vec{u}^o,\vec{v})\delta_o - (\vec{u}^N,\vec{v})\delta_T \ ,$$

whose Fourier transform with respect to t is :

$$2i\pi\tau(\hat{\vec{u}}_k(\tau),\vec{v}) + < \hat{\vec{g}}_k(\tau),\vec{v} > = < \hat{\vec{f}}_k(\tau),\vec{v} > + (\vec{u}^o,\vec{v}) - (\vec{u}^N,\vec{v})\ e^{-2i\pi\tau T} \ .$$

The imaginary part of this equation is bounded as follows :

$$2\pi|\tau|\ |\hat{\vec{u}}_k(\tau)|^2 \leqslant (\|\hat{\vec{g}}_k(\tau)\|_{\star} + \|\hat{\vec{f}}_k(\tau)\|_{\star})\|\hat{\vec{u}}_k(\tau)\| + (|\vec{u}^o| + |\vec{u}^N|)|\hat{\vec{u}}_k(\tau)| \ . \tag{2.10}$$

Now, it can be readily seen that

$$\|\hat{\vec{g}}_k(\tau)\|_{\star} \leqslant \sum_{n=0}^{N-1} \int_{t_n}^{t_{n-1}} \|\vec{g}_k(t)\|_{\star}dt \leqslant k \sum_{n=0}^{N-1} \{\nu \|\vec{u}^{n+1}\| + C_1 \|\vec{u}^n\|\ \|\vec{u}^{n+1}\|\} \ .$$

Hence, by virtue of (2.6) ,

$$\|\vec{g}_k(\tau)\|_\star \leqslant C_2 \quad .$$

Likewise, $\quad \|\vec{f}_k(\tau)\|_\star \leqslant k \sum_{n=0}^{N-1} \|\vec{f}^{n-1}\|_\star \leqslant \|\vec{f}\|_{L^2(0,T\,;\,V')} \; .$

Therefore (2.10) reduces to :

$$|\tau|\|\hat{\vec{u}}_k(\tau)|^2 \leqslant C_3 \|\hat{\vec{u}}_k(\tau)\| + C_4|\hat{\vec{u}}_k(\tau)| \qquad \forall\, \tau \in \mathbb{R} \,.$$

Like in the end of Lemma 1.7, this implies that

$$\int_{-\infty}^{+\infty} |\tau|^{2\gamma}\, |\hat{\vec{u}}_k(\tau)|^2 d\tau \leqslant C_5 \text{ for some } \gamma \text{ such that } 0 < \gamma < \frac{1}{4} \; . \; ■$$

THEOREM 2.1.

Suppose that the initial value $\vec{u}_o \in V$ and choose $\vec{u}^o = \vec{u}_o$. Then

$$\text{weak} \lim_{k \to 0} \vec{u}_k = \vec{u} \;\; \underline{\text{in}} \;\; L^2(0,T\,;\,V) \;,$$

$$\text{weak}\star \lim_{k \to 0} \vec{u}_k = \vec{u} \;\; \underline{\text{in}} \;\; L^\infty(0,T\,;\,H) \;,$$

$$\lim_{k \to 0} \vec{u}_k = \vec{u} \;\; \underline{\text{in}} \;\; L^2(0,T\,;\,H) \;,$$

where $\vec{u}$ is a solution of problem (P).

PROOF.

Lemma 2.3 implies that there exists a subsequence of $(\vec{u}_k)$, also called $\vec{u}_k$ for convenience, such that :

$$\text{weak} \lim_{k \to 0} \vec{u}_k = \vec{u} \text{ in } L^2(0,T\,;\,V) \;,\; \text{weak}\star \lim_{k \to 0} \vec{u}_k = \vec{u} \text{ in } L^\infty(0,T\,;\,H)$$

and $\displaystyle\lim_{k \to 0} \vec{u}_k = \vec{u}$ in $L^2(0,T\,;\,H)$. It remains to prove that $\vec{u}$ satisfies (2.1). For this, it is convenient to introduce the following step function :

$$\vec{w}_k(t) = \vec{u}_k(t_{n+1}) \text{ on }]\,t_n,t_{n+1}] \quad \text{for } -1 \leqslant n \leqslant N-1 \;.$$

Clearly, (2.6) and (2.5) imply that

$$\text{weak} \lim_{k \to 0} \vec{w}_k = \vec{w} \text{ in } L^2(0,T\,;\,V) \text{ and weak} \star \lim_{k \to 0} \vec{w}_k = \vec{w} \text{ in } L^\infty(0,T\,;\,H).$$

Furthermore (2.7) implies that :

$$\lim_{k \to 0} (\vec{w}_k - \vec{u}_k) = 0 \quad \text{in} \quad L^2(0,T\,;\,H)\,.$$

Hence, $\vec{w} = \vec{u}$ and $\lim_{k \to 0} \vec{w}_k = \vec{u}$ in $L^2(0,T\,;\,H)$.

Now, it suffices to rewrite (2.2) with these functions :

$$\frac{d}{dt}(\vec{u}_k(t),\vec{v}) + a_o(\vec{w}_k(t),\vec{v}) + a_1(\vec{w}_k(t-k)\,;\,\vec{w}_k(t),\vec{v}) = \langle \vec{f}_k(t),\vec{v} \rangle$$

and to pass to the limit like in Theorem 1.4 in order to recover (2.1). ∎

Let us examine now the discretization error when the exact solution $\vec{u}$ is sufficiently smooth. More precisely, we assume that

$$\text{(2.11)} \qquad \frac{d\vec{u}}{dt} \in L^2(0,T\,;\,V) \quad , \quad \frac{d^2\vec{u}}{dt^2} \in L^2(0,T\,;\,V')\,.$$

In view of Lemma 1.1 and Theorem 1.1, this implies that :

$$\vec{u} \in \mathcal{C}^o([\,0,T\,]\,;\,V) \quad \text{and} \quad \frac{d\vec{u}}{dt} \in \mathcal{C}^o([\,0,T\,]\,;\,H)\,.$$

As a consequence,

$$t \longmapsto a(\vec{u}(t)\,;\,\vec{u}(t),\vec{v}) \in \mathcal{C}^o([\,0,T\,])\,,$$

$$\frac{d}{dt}(\vec{u}(t),\vec{v}) = (\frac{d}{dt}\,\vec{u}(t),\vec{v}) \in \mathcal{C}^o([\,0,T\,]).$$

Hence the right-hand side $t \longmapsto \langle \vec{f}(t),\vec{v} \rangle \in \mathcal{C}^o([\,0,T\,])$ and (2.1) reads as follows :

$$\text{(2.12)} \qquad (\frac{d\vec{u}}{dt}(t),\vec{v}) + a(\vec{u}(t)\,;\,\vec{u}(t),\vec{v}) = \langle \vec{f}(t),\vec{v} \rangle \quad \forall\, t \in [\,0,T\,]\,,\ \forall\, \vec{v} \in \mathbf{V}\,.$$

Now, we define the discretization error $\vec{e}^{\,n} \in \mathbf{V}$ and the truncation (or consistency) error $\vec{\varepsilon}^{\,n} \in V'$ by :

$$\text{(2.13)} \qquad \vec{e}^{\,n} = \vec{u}(t_n) - \vec{u}^{\,n} \qquad \text{for} \quad 0 \leqslant n \leqslant N\,,$$

$$\text{(2.14)} \qquad \langle \vec{\varepsilon}^{\,n},\vec{v} \rangle = \frac{1}{k}(\vec{u}(t_{n+1}) - \vec{u}(t_n),\vec{v}) + a(\vec{u}(t_n)\,;\,\vec{u}(t_{n+1}),\vec{v}) - \langle \vec{f}(t_{n+1}),\vec{v} \rangle$$

$$\forall\, \vec{v} \in \mathbf{V} \quad , \ 0 \leqslant n \leqslant N-1\,.$$

Following the classical argument used in ordinary differential equations, we Propose to show that the scheme (2.2) is stable and that the consistency error is of order one. But beforehand, let us establish the following discrete analogue of Lemma 1.8.

LEMMA 2.4.

Let (a_n), (b_n) and (c_n) be three sequences of positive real numbers such that (c_n) is monotonically increasing and

$$a_n + b_n \leqslant c_n + \lambda \sum_{m=0}^{n-1} a_m \quad \text{for } n \geqslant 1 \text{ and } \lambda > 0, \tag{2.15}$$

with

$$a_o + b_o \leqslant c_o .$$

Then, these sequences also satisfy

$$a_n + b_n \leqslant c_n \exp(\lambda n) \quad \text{for } n \geqslant 0 .$$

PROOF.

Let us show by induction that (2.15) implies :

$$a_n + b_n \leqslant c_n (1+\lambda)^n \quad \text{for } n \geqslant 0 . \tag{2.16}$$

This is obviously true for $n = 0$. Next, suppose it is true for all $n \leqslant n_o$ and let $n = n_o+1$. By (2.15), the induction hypothesis and the monotonicity of (c_n), we get :

$$a_{n_o+1} + b_{n_o+1} \leqslant c_{n_o+1}\Big(1 + \lambda \sum_{m=0}^{n_o} (1+\lambda)^m\Big) = c_{n_o+1}(1+\lambda)^{n_o+1},$$

thus proving (2.16). Then, the lemma follows from the fact that

$$(1+\lambda)^n \leqslant \exp(n\lambda) . \quad \blacksquare$$

LEMMA 2.5.

Assume that $\vec{u}_o \in V$. Then, if $\vec{u}^o = \vec{u}_o$, the following stability criterion holds :

$$|\vec{e}^n|^2 + \sum_{m=0}^{n-1} |\vec{e}^{m+1} - \vec{e}^m|^2 + \nu k \sum_{m=1}^{n} \|\vec{e}^m\|^2 \leqslant \frac{2}{\nu}\Big(k \sum_{m=0}^{n-1} \|\vec{\varepsilon}^m\|_*^2\Big)\exp(ct_n), \tag{2.17}$$

$$\text{for } 1 \leqslant n \leqslant \mathbb{N} .$$

PROOF.

First, consider the expression :

$$E_n = \frac{1}{k}(\vec{e}^{n+1} - \vec{e}^n, \vec{v}) + a(\vec{u}^n; \vec{e}^{n+1}, \vec{v}) \quad \text{for} \quad 0 \leq n \leq N-1 .$$

Using (2.13) and (2.2) , we find :

$$E_n = \frac{1}{k}(\vec{u}(t_{n+1}) - \vec{u}(t_n), \vec{v}) + a(\vec{u}^n ; \vec{u}(t_{n+1}), \vec{v}) - < \vec{f}(t_{n+1}), \vec{v} > .$$

Therefore, in view of (2.14) the error $\vec{e}^n$ satisfies the equation :

$$\text{(2.18)} \qquad \frac{1}{k}(\vec{e}^{n+1} - \vec{e}^n, \vec{v}) + a(\vec{u}^n ; \vec{e}^{n+1}, \vec{v}) = < \vec{\varepsilon}^n, \vec{v} > - a_1(\vec{e}^n ; \vec{u}(t_{n+1}), \vec{v})$$
$$\forall \vec{v} \in \mathbf{V} , \quad 0 \leq n \leq N-1 .$$

Next, take $\vec{v} = \vec{e}^{n+1}$ in (2.18) and use the identity (2.8).

We obtain :

$$\text{(2.19)} \qquad |\vec{e}^{n+1}|^2 - |\vec{e}^n|^2 + |\vec{e}^{n+1} - \vec{e}^n|^2 + 2k\nu \|\vec{e}^{n+1}\|^2 = 2k < \vec{\varepsilon}^n, \vec{e}^{n+1} >$$
$$- 2ka_1(\vec{e}^n ; \vec{u}(t_{n+1}), \vec{e}^{n+1}) .$$

The right-hand side is estimated as follows :

$$\text{(2.20)} \qquad 2| < \vec{\varepsilon}^n, \vec{e}^{n+1} > | \leq 2 \|\vec{\varepsilon}^n\|_\star \|\vec{e}^{n+1}\| \leq \frac{\nu}{2} \|\vec{e}^{n+1}\|^2 + \frac{2}{\nu} \|\vec{\varepsilon}^n\|_\star^2 ,$$

$$|a_1(\vec{e}^n ; \vec{u}(t_{n+1}), \vec{e}^{n+1})| \leq C_1 \|\vec{e}^n\|_{o,3,\Omega} \|\vec{u}(t_{n+1})\| \|\vec{e}^{n+1}\|_{o,6,\Omega}$$
$$\leq C_2 \|\vec{e}^n\|_{o,3,\Omega} \|\vec{e}^{n+1}\| ,$$

by Sobolev's imbedding theorem, provided the dimension does not exceed 3 . But in that case, all functions φ of $H^1_o(\Omega)$ satisfy the inequality :

$$\|\varphi\|_{o,3,\Omega} \leq \|\varphi\|^{1/2}_{o,6,\Omega} \|\varphi\|^{1/2}_{o,2,\Omega} \leq C_3 |\varphi|^{1/2}_{1,\Omega} \|\varphi\|^{1/2}_{o,\Omega} \quad \text{(cf.(1.22))}.$$

Therefore

$$2|a_1(\vec{e}^n ; \vec{u}(t_{n+1}), \vec{e}^{n+1})| \leq 2 C_4 \|\vec{e}^{n+1}\| \|\vec{e}^n\|^{1/2} |\vec{e}^n|^{1/2} .$$
$$\leq C_4\{\varepsilon \|\vec{e}^{n+1}\|^2 + \frac{1}{\varepsilon} \|\vec{e}^n\| |\vec{e}^n|\}$$
$$\leq C_4\{\varepsilon \|\vec{e}^{n+1}\|^2 + \frac{1}{2\varepsilon}(\delta \|\vec{e}^n\|^2 + \frac{1}{\delta}|\vec{e}^n|^2)\} ,$$

where $\varepsilon > 0$ and $\delta > 0$ are arbitrary. Hence we have the estimate :

$$(2.21)\qquad 2|a_1(\vec{e}^n\ ;\ \vec{u}(t_{n+1}),\vec{e}^{n+1})| \leq \frac{\nu}{4}\,(\|\vec{e}^{n+1}\|^2 + \|\vec{e}^n\|^2) + C_5(\nu)|\vec{e}^n|^2 .$$

Let us substitute (2.20) and (2.21) in (2.19) :

$$|\vec{e}^{n+1}|^2 - |\vec{e}^n|^2 + |\vec{e}^{n+1} - \vec{e}^n|^2 + 2k\nu\,\|\vec{e}^{n+1}\|^2$$
$$\leq k(\frac{3\nu}{4}\|\vec{e}^{n+1}\|^2 + \frac{\nu}{4}\|\vec{e}^n\|^2 + \frac{2}{\nu}\|\vec{\varepsilon}^n\|_\star^2 + C_5(\nu)|\vec{e}^n|^2).$$

Then by summing from 0 to $m-1$ and using the fact that $\vec{e}^o = \vec{0}$,

we obtain :

$$|\vec{e}^m|^2 + \sum_{n=0}^{m-1}|\vec{e}^{n+1} - \vec{e}^n|^2 + k\nu\sum_{n=1}^{m}\|\vec{e}^n\|^2 \leq \frac{2k}{\nu}\sum_{n=0}^{m-1}\|\vec{\varepsilon}^n\|_\star^2 + kC_5(\nu)\sum_{n=0}^{m-1}|\vec{e}^n|^2 .$$

It suffices to apply Lemma 2.4 with $a_n = |\vec{e}^n|^2$, $\lambda = kC_5(\nu), c_o = 0$,

$$c_n = \frac{2k}{\nu}\sum_{m=0}^{n-1}\|\vec{\varepsilon}^m\|_\star^2 ,\ b_o = 0 ,\ b_n = \sum_{m=0}^{n-1}|\vec{e}^{m+1} - \vec{e}^m|^2 + k\nu\sum_{m=1}^{n}\|\vec{e}^m\|^2 ,$$

in order to derive (2.17) with $c = kC_5(\nu)$. ■

<u>LEMMA 2.6</u>.

<u>Suppose that</u> $\vec{u}$ <u>satisfies the regularity conditions (2.11) and that</u> $\vec{u}^o = \vec{u}_o \in \mathbf{V}$. <u>Then the truncation error is of the order of</u> k , <u>i.e</u>.

$$(2.22)\qquad (k\sum_{m=0}^{n-1}\|\vec{\varepsilon}^m\|_\star^2)^{1/2} \leq Ck \quad \underline{\text{for}} \quad 1 \leq n \leq N .$$

<u>PROOF</u>.

By combining (2.12) and (2.14) , we get :

$$(2.23)\ <\vec{\varepsilon}^n,\vec{v}> = \frac{1}{k}(\vec{u}(t_{n+1}) - \vec{u}(t_n),\vec{v}) - (\frac{d}{dt}\vec{u}(t_{n+1}),\vec{v}) - a_1(\vec{u}(t_{n+1}) - \vec{u}(t_n);\vec{u}(t_{n+1}),\vec{v}) .$$

By virtue of (1.1), and Taylor's expansion with integral remainder, namely

$$\frac{1}{k}(\vec{u}(t_{n+1}) - \vec{u}(t_n),\vec{v}) = (\frac{d\vec{u}}{dt}(t_{n+1}),\vec{v}) + \frac{1}{k}\int_{t_n}^{t_{n+1}}(t-t_{n+1}) < \frac{d^2\vec{u}}{dt^2}(t),\vec{v} > dt,$$

(2.23) can also be written :

$$<\vec{\varepsilon}^n,\vec{v}> = \frac{1}{k}\int_{t_n}^{t_{n+1}}(t-t_{n+1}) < \frac{d^2\vec{u}}{dt^2}(t),\vec{v} > dt - a_1(\int_{t_n}^{t_{n+1}}\frac{d\vec{u}}{dt}(t)\,dt\ ;\ \vec{u}(t_{n+1}),\vec{v}).$$

Hence

$$\|\vec{\varepsilon}^n\|_\star \leq \frac{1}{k}\int_{t_n}^{t_{n+1}} (t_{n+1}-t)\|\frac{d^2\vec{u}(t)}{dt^2}\|_\star dt + C_1 \left\|\int_{t_n}^{t_{n+1}} \frac{d\vec{u}(t)}{dt}\, dt\right\| \|\vec{u}(t_{n+1})\| .$$

Therefore,

$$\|\vec{\varepsilon}^n\|_\star^2 \leq C_2 k \int_{t_n}^{t_{n+1}} \{\|\frac{d^2\vec{u}(t)}{dt^2}\|_\star^2 + \|\frac{d\vec{u}(t)}{dt}\|^2\}dt$$

and

$$k\sum_{m=0}^{n-1} \|\vec{\varepsilon}^m\|_\star^2 \leq C_2 k^2\left[\|\frac{d^2\vec{u}}{dt^2}\|^2_{L^2(0,T\,;\,V')} + \|\frac{d\vec{u}}{dt}\|^2_{L^2(0,T\,;\,V)}\right]. \quad \blacksquare$$

Combining Lemmas 2.5 and 2.6, we derive immediately the next error bound which shows that the method (2.2) is *of order one.*

THEOREM 2.2

Under the hypotheses of Lemma 2.6, there exists a constant $C(\vec{u}) > 0$ such that

$$\max_{0\leq n\leq N} |\vec{u}(t_n) - \vec{u}^n| + (k\sum_{n=1}^{N} \|\vec{u}(t_n) - \vec{u}^n\|^2)^{1/2} \leq C(\vec{u})k .$$

§ 3 - Semi-discretization with a multistep method

The drawback of the one-step, semi-discrete method analyzed in the preceding paragraph is that it is only a first-order method. It is well known with ordinary differential equations that when more accuracy is required, it is worthwhile to resort to multistep methods. As a consequence, we propose here to adapt some multistep methods to the semi-discretization of problem (P).

3.1. Generalities about multistep methods.

Consider the initial value problem

$$(3.1) \qquad y' = f(t,y) \quad , \quad t \in [0,T] \quad , \quad y(0) = y_o \quad ,$$

and assume that we know q approximate values of $y(t)$: $y_o, y_1, \ldots, y_{q-1}$, at the points $t_o, \dots, t_{q-1}$ respectively.

DEFINITION 3.1.

A q-step method for solving (3.1) consists in finding a sequence (y_n), $q \leq n \leq N$, defined by :

$$(3.2) \qquad \sum_{i=0}^{q} \alpha_i y_{n+i} = k \sum_{i=0}^{q} \beta_i f(t_{n+i}, y_{n+i}) \quad \text{for} \quad 0 \leq n \leq N-q ,$$

where α_i and β_i are real parameters satisfying $\alpha_q \neq 0$ and $|\alpha_o| + |\beta_o| \neq 0$.

Of course, the sequence (y_n) is unaffected if we multiply (3.2) by some factor and therefore, we decide to normalize (3.2) by the condition :

$$\sum_{i=0}^{q} \beta_i = 0 .$$

It is convenient to associate with the multistep method (3.2) a pair of polynomials (ρ,σ) of degree q defined by :

$$(3.3) \qquad \rho(\zeta) = \sum_{i=0}^{q} \alpha_i \zeta^i , \quad \sigma(\zeta) = \sum_{i=0}^{q} \beta_i \zeta^i .$$

Since there is an obvious one-to-one correspondence between (3.2) and (3.3), the multistep method (3.2) is often called a (ρ,σ)-method . Moreover, the definitions of order and stability are stated in terms of these polynomials.

DEFINITION 3.2.

The (ρ,σ)-method (3.2) is said to be of *order* p if p is the largest integer such that

$$(3.4) \qquad \frac{1}{k} \sum_{i=0}^{q} \{\alpha_i z(t+ik) - k\beta_i z'(t+ik)\} = O(k^p) \quad \text{as } k \text{ tends to zero} ,$$ for all sufficiently smooth functions z .

By means of Taylor's expansion, it is easy to check that the (ρ,σ)-method (3.2) is of order p if and only if

$$(3.5) \qquad c_\ell = 0 \qquad 0 \leq \ell \leq p ,$$

where the coefficients c_ℓ are defined by :

(3.6) $$c_o = \sum_{i=0}^{q} \alpha_i \quad , \quad c_\ell = \frac{1}{(\ell-1)!} \sum_{i=0}^{q} \{\frac{1}{\ell} i^\ell \alpha_i - i^{\ell-1}\beta_i\} \text{ for } 1 \leqslant \ell \leqslant p \ ,$$

(with the convention $0^o = 0! = 1$).

In order to introduce the notion of stability of a multistep method, consider the linear differential equation with constant coefficients :

(3.7) $$y' = -\lambda y \quad , \quad \text{with} \quad \lambda \in \mathbb{C} \quad , \quad y(0) = y_o \quad ,$$

whose solution is $y(t) = y_o \exp(-\lambda t)$. Observe that y is bounded, i.e. $\sup_{t \geqslant 0} |y(t)| < +\infty$ iff $\mathcal{Re}\, \lambda \geqslant 0$. Then, it is reasonable to expect that the approximate solution defined by (3.2) should also be bounded for the same range of λ. This property, called A-stability, is defined as follows.

DEFINITION 3.3.

1) The (ρ,σ)-method (3.2) is called A-*stable* if the sequence $(y_n(Z))$ defined by : $y_o, y_1, \ldots, y_{q-1}$ arbitrary ,

(3.8) $$\sum_{i=0}^{q} (\alpha_i + Z\beta_i) y_{n+i}(Z) = 0 \quad \text{for} \quad n \geqslant 0$$

satisfies

$$\sup_{n \geqslant 0} |y_n(Z)| < +\infty \qquad \forall\, \mathcal{Re}\, Z \geqslant 0 \ .$$

2) Furthermore, it is called *strongly A-stable* if it is A-stable and if, in addition, the roots σ_i of σ satisfy $|\sigma_i| < 1$ for $1 \leqslant i \leqslant q$.

Note that (3.8) is simply obtained by applying (3.2) to (3.7) and setting $Z = \lambda k$.

As far as the order of A-stable methods is concerned, Dahlquist [19] has proved the following crucial, though negative, result.

THEOREM 3.1.

There exists no A-stable (ρ,σ)-method whose order exceeds 2 .

In view of this result, and since it is vital that a multistep method be A-stable, the best we can do is to derive one-step or two-step, A-stable methods of order two. They are given in the examples below.

Example 3.1.

All one-step (ρ,σ)-methods of order ≥ 1 have the form :

$$(3.9) \qquad \rho(\zeta) = \zeta - 1 \quad , \quad \sigma(\zeta) = \theta\zeta + 1-\theta \quad \text{for} \quad \theta \in \mathbb{R} \ ;$$

they are called θ-methods . A θ-method is of order one for all $\theta \neq \frac{1}{2}$ and of order two when $\theta = \frac{1}{2}$, in which case it corresponds to the trapezoidal rule. It is easy to check from Definition 3.3 that it is A-stable when $\theta \geq \frac{1}{2}$ and strongly A-stable when $\theta > \frac{1}{2}$. Hence, for $\theta = \frac{1}{2}$, it is at the same time A-stable and of order 2 . ■

Example 3.2.

All two-step (ρ,σ)-methods of order ≥ 2 have the form:

$$\rho(\zeta) = \alpha_2\zeta^2 + \alpha_1\zeta + \alpha_o \quad , \quad \sigma(\zeta) = \beta_2\zeta^2 + \beta_1\zeta + \beta_o$$

with

$$(3.10) \quad \begin{cases} \alpha_o = -1 + \alpha_2 \quad , \quad \alpha_1 = 1 - 2\alpha_2 \quad , \\ \beta_o = \frac{1}{2} - \alpha_2 + \beta_2 \ , \ \beta_1 = \frac{1}{2} + \alpha_2 - 2\beta_2 \quad . \end{cases}$$

Such methods are A-stable for $\alpha_2 \geq \frac{1}{2}$ and $\beta_2 > \frac{\alpha_2}{2}$ and strongly A-stable for $\alpha_2 > \frac{1}{2}$ and $\beta_2 > \frac{\alpha_2}{2}$. ■

3.2. Multistep methods for solving the Navier-Stokes problem.

From now on, we assume that the right-hand side $\vec{f}$ of problem (2.1) belongs to $\mathcal{C}^o([0,T] \ ; (H^{-1}(\Omega))^n)$. Then, the straightforward application of the scheme

(3.2) to this problem consists in finding $\vec{u}^q$, $\vec{u}^{q+1}$,..., such that

$$\text{(3.11)} \quad \begin{cases} \frac{1}{k} \sum_{i=0}^{q} \alpha_i(\vec{u}^{n+i},\vec{v}) + \sum_{i=0}^{q} \beta_i\{a(\vec{u}^{n+i} ; \vec{u}^{n+i},\vec{v}) - < \vec{f}(t_{n+i}),\vec{v} >\} = 0 \\ \qquad\qquad\qquad \forall\, \vec{v} \in \mathbf{V} ,\ \forall\, n \geqslant 0 , \\ \text{starting from } q \text{ functions : } \vec{u}^o ,..., \vec{u}^{q-1} \text{, given in } \mathbf{V} . \end{cases}$$

Unfortunately, the practical solution of (3.11) is much too expensive, because it is a non linear equation for the unknown $\vec{u}^{n+q}$. Therefore, we wish to linearize this scheme without decreasing its order. The simplest way to achieve this is to replace the non-linear expression

$$\text{(3.12)} \qquad \sum_{i=0}^{q} \beta_i\{a(\vec{u}^{n+i} ; \vec{u}^{n+i} , \vec{v}) - < \vec{f}(t_{n+i}),\vec{v} >\}$$

by

$$\sum_{i=0}^{q} \beta_i\{a(\vec{u}^{\bar{n}} ; \vec{u}^{n+i},\vec{v}) - < \vec{f}(t_{\bar{n}}),\vec{v} > ,$$

where $\vec{u}^{\bar{n}}$ is a linear combination of $\vec{u}^n ,..., \vec{u}^{n+q-1}$ and $t_{\bar{n}} \in [t_n,t_{n+q}]$, and both are chosen so as to obtain at least a second order scheme,as this is the best order of an A-stable method.

The choice of $t_{\bar{n}}$ is suggested by the following considerations. We observe that if φ is sufficiently smooth, then by virtue of the normalizing condition $\sum_{i=0}^{q} \beta_i = 1$, we can write that

$$\text{(3.13)} \qquad \sum_{i=0}^{q} \beta_i\, \varphi(t_{n+i}) = \varphi\,(\sum_{i=0}^{q} \beta_i t_{n+i}) + O(k^2) \quad \text{as} \quad k \to 0 .$$

Thus, if we set

$$\text{(3.14)} \qquad t_{\bar{n}} = \sum_{i=0}^{q} \beta_i t_{n+i} ,$$

then we can at first replace (3.12) by :

$$\text{(3.15)} \qquad \sum_{i=0}^{q} \beta_i a(\sum_{j=0}^{q} \beta_j \vec{u}^{n+j} ; \vec{u}^{n+i},\vec{v}) - < \vec{f}(t_{\bar{n}}),\vec{v} > ,$$

without decreasing the order of the scheme.

In order to choose $\vec{u}^{\bar{n}}$, let $\tilde{\varphi}$ be the polynomial of degree $\leq p-1$ which interpolates φ at the point $t_{n+q-p}, \ldots, t_{n+q-1}$. We have :

$$\varphi(t_{n+q}) = \tilde{\varphi}(t_{n+q}) + O(k^p) = \sum_{i=q-p}^{q-1} c_i\, \varphi(t_{n+i}) + O(k^p) \quad \text{as} \quad k \to 0 ,$$

where the coefficients c_i are independent of φ. Therefore, we can write

$$(3.16) \qquad \sum_{i=0}^{q} \beta_i\, \varphi(t_{n+i}) = \sum_{i=0}^{q-1} \beta_i \varphi(t_{n+i}) + \beta_q \sum_{i=q-p}^{q-1} c_i\, \varphi(t_{n+i}) + O(k^p) .$$

Thus, if $p \geq q$ and if we set

$$(3.17) \qquad \gamma_i = \begin{cases} c_i \beta_q & \text{for} \quad q-p \leq i \leq -1 \\ \beta_i + c_i \beta_q & \text{for} \quad 0 \leq i \leq q-1 , \end{cases}$$

then we can replace (3.11) by the following scheme : find $\vec{u}^q, \vec{u}^{q+1}, \ldots, \vec{u}^N$:

$$(3.18) \begin{cases} \dfrac{1}{k} \displaystyle\sum_{i=0}^{q} \alpha_i (\vec{u}^{n+i}, \vec{v}) + \sum_{i=0}^{q} \beta_i a(\vec{u}^{\bar{n}} ; \vec{u}^{n+i}, \vec{v}) = \langle \vec{f}(t_{\bar{n}}), \vec{v} \rangle \ \forall \vec{v} \in \mathbf{V} \\ \qquad\qquad\qquad\qquad\qquad\qquad\qquad\qquad 0 \leq n \leq N-q , \\ \text{where} \\ \qquad \vec{u}^{\bar{n}} = \displaystyle\sum_{j=q-p}^{q-1} \gamma_j \vec{u}^{n+j} , \end{cases}$$

with γ_j defined by (3.17) and $t_{\bar{n}}$ by (3.14) and where the starting values $\vec{u}^{q-p}, \ldots, \vec{u}^{q-1}$ are given in $\mathbf{V}$.

It follows from (3.13) and (3.16) that the order of the method (3.18) is at least two, provided the original method (3.11) is at least of order two.

Note that this technique for linearization is in fact an extrapolation since the value of $\vec{u}^{n+q}$ is predicted by a polynomial extrapolation.

Let us adapt this linearization process to the examples of section 1.

<u>Example 3.1</u>.

When $\theta \neq \frac{1}{2}$, the θ-method is of the first order. Therefore, it suffices to take $p = 1$ and $\vec{u}^{\bar{n}} = \vec{u}^n$. In particular, when $\theta = 1$, it yields the one-step method of paragraph 1. When $\theta = \frac{1}{2}$, we must take $p = 2$; then (3.17) gives :

$$\vec{u}^{\bar{n}} = \frac{3}{2}\vec{u}^{n} - \frac{1}{2}\vec{u}^{n-1} ,$$

a choice which has been first introduced by Douglas and Dupont [21] . ■

Example 3.2.

Consider the A-stable two-step methods of order two given by (3.10). Here, we can take $p = 2$ and (3.17) yields the following expression for $\vec{u}^{\bar{n}}$:

$$\vec{u}^{\bar{n}} = (\frac{1}{2} - \alpha_2)\vec{u}^{n} + (\frac{1}{2} + \alpha_2)\vec{u}^{n+1} . \tag{3.19}$$ ■

3.3. Convergence of a family of two-step methods

As mentioned in section 3.1, we are primarily interested in one-step or two-step, A-stable methods of order two. With the linearization of the last section, this choice leads to the methods described in the Examples 3.1 and 3.2. Now, the analysis of the one-step methods of Example 3.1 is entirely similar to that developed in § 2 . Therefore, the remainder of this paragraph is devoted to the study of the two-step methods given by the Example 3.2 , namely :

$$\text{(3.20)} \begin{cases} \frac{1}{k}\{(\alpha_2-1)(\vec{u}^{n},\vec{v}) + (1-2\alpha_2)(\vec{u}^{n+1},\vec{v}) + \alpha_2(\vec{u}^{n+2},\vec{v})\} \\ + (\frac{1}{2} - \alpha_2 + \beta_2)a(\vec{u}^{\bar{n}} ; \vec{u}^{n},\vec{v}) + (\frac{1}{2} + \alpha_2 - 2\beta_2)a(\vec{u}^{\bar{n}} ; \vec{u}^{n+1},\vec{v}) \\ + \beta_2 a(\vec{u}^{\bar{n}} ; \vec{u}^{n+2},\vec{v}) = \langle \vec{f}(t_{\bar{n}}),\vec{v} \rangle \qquad \forall \vec{v} \in \mathbf{V} , \\ \text{where} \\ t_{\bar{n}} = t_n + (\alpha_2 + \frac{1}{2})k , \\ \vec{u}^{\bar{n}} \text{ is given by (3.19)} , \quad \alpha_2 \geq \frac{1}{2} \text{ and } \beta_2 > \frac{\alpha_2}{2} . \end{cases}$$

LEMMA 3.1.

For each pair of starting values $\vec{u}^{\circ}$ and $\vec{u}^{1} \in \mathbf{V}$, the scheme (3.20) defines a unique sequence $(\vec{u}^{n}) \subset \mathbf{V}$.

PROOF.

The function $\vec{u}^{n+2}$ is the solution in $\mathbb{V}$ of the linear boundary-value problem :

$$\alpha_2(\vec{u}^{n+2},\vec{v}) + k\beta_2 a(\vec{u}^{\bar{n}} ; \vec{u}^{n+2},\vec{v}) = \langle \vec{\ell},\vec{v} \rangle \qquad \forall\, \vec{v} \in \mathbb{V} ,$$

where $\vec{\ell}$ is a known element of $\mathbb{V}'$. But this is an elliptic problem corresponding to a continuous, bilinear and elliptic form on $\mathbb{V}$ since, by hypothesis , $\alpha_2 > 0$ and $\beta_2 > 0$. ∎

Like in § 2, let us introduce the truncation error $\vec{\varepsilon}^n \in \mathbb{V}'$, defined by :

$$(3.21)\quad \begin{cases} \langle \vec{\varepsilon}^n,\vec{v} \rangle = \frac{1}{k} \sum_{i=0}^{2} \alpha_i(\vec{u}(t_{n+i}),\vec{v}) + \sum_{i=0}^{2} \beta_i a(\sum_{j=0}^{1} \gamma_j \vec{u}(t_{n+j}) ; \vec{u}(t_{n+i}) , \vec{v}) \\ \quad - \langle \vec{f}(t_{\bar{n}}),\vec{v} \rangle \qquad \forall\, \vec{v} \in \mathbb{V} , \end{cases}$$

and the pointwise error $\vec{e}^n \in \mathbb{V}$:

$$\vec{e}^n = \vec{u}(t_n) - \vec{u}^n \qquad \text{for} \quad 0 \leq n \leq N .$$

The next lemma gives the expected estimate for $\vec{\varepsilon}^n$, since according to the preceding section, the scheme (3.20) is of order two.

LEMMA 3.2.

Assume that the solution $\vec{u}$ of the Navier-Stokes equations has the following regularity :

$$(3.22)\qquad \vec{u} , \frac{d\vec{u}}{dt} , \frac{d^2\vec{u}}{dt^2} \in L^2(0,T ; \mathbb{V}) , \frac{d^3\vec{u}}{dt^3} \in L^2(0,T ; \mathbb{V}') .$$

Then the truncation error is bounded as follows :

$$(3.23)\qquad (k \sum_{m=0}^{n-2} \|\vec{\varepsilon}^m\|_\star^2)^{1/2} \leq ck^2 \Big[\int_0^{t_n} (\|\frac{d^2\vec{u}}{dt^2}(t)\|^2 + \|\frac{d^3\vec{u}}{dt^3}(t)\|_\star^2)dt\Big]^{1/2} ,$$

where the constant c depends upon $\vec{u}$ only.

PROOF.

Let us expand $\vec{u}(t_n)$, $\vec{u}(t_{n+1})$ and $\vec{u}(t_{n+2})$ about the point $t_{\bar{n}}$ by Taylor's formula with integral remainder . We obtain :

$$\frac{1}{k}\sum_{i=0}^{2}\alpha_i\vec{u}(t_{n+i}) = \frac{d\vec{u}}{dt}(t_{\bar{n}}) + k\int_{t_n}^{t_{n+2}} K_1(t)\frac{d^3\vec{u}}{dt^3}(t)dt,$$

$$\sum_{i=0}^{1}\gamma_i\vec{u}(t_{n+i}) = \vec{u}(t_{\bar{n}}) + k\int_{t_n}^{t_{n+2}} K_2(t)\frac{d^2\vec{u}}{dt^2}(t)dt,$$

$$\sum_{i=0}^{2}\beta_i\vec{u}(t_{n+i}) = \vec{u}(t_{\bar{n}}) + k\int_{t_n}^{t_{n+2}} K_3(t)\frac{d^2\vec{u}}{dt^2}(t)dt,$$

where the kernels $K_1(t)$, $K_2(t)$ and $K_3(t)$ are bounded by a constant c_1 independent of k and $\vec{u}$. Next, let us substitute these expansions in (3.21) and observe that

$$\sum_{i=0}^{2}\beta_i a\left[\vec{u}(t_{\bar{n}}) + k\int_{t_n}^{t_{n+2}} K_2(t)\frac{d^2\vec{u}}{dt^2}(t)\,dt\ ;\ \vec{u}(t_{n+i}),\vec{v}\right]$$

$$= a(\vec{u}(t_{\bar{n}})\ ;\ \vec{u}(t_{\bar{n}}),\vec{v}) + k\int_{t_n}^{t_{n+2}} K_3(t)a(\vec{u}(t_{\bar{n}})\ ;\ \frac{d^2\vec{u}}{dt^2}(t),\vec{v})dt$$

$$+ k\int_{t_n}^{t_{n+2}} K_2(t)a_1(\frac{d^2\vec{u}}{dt^2}(t)\ ;\ \vec{u}(t_{\bar{n}}),\vec{v})dt$$

$$+ k^2\, a_1\left[\int_{t_n}^{t_{n+2}} K_2(t)\frac{d^2\vec{u}}{dt^2}(t)\,dt\ ;\ \int_{t_n}^{t_{n+2}} K_3(t)\frac{d^2\vec{u}}{dt^2}(t)\,dt,\vec{v}\right].$$

Then by using (2.12) at $t = t_{\bar{n}}$, we find

$$\langle \vec{\varepsilon}^{\,n},\vec{v}\rangle = k\int_{t_n}^{t_{n+2}}\{K_1(t)\langle\frac{d^3\vec{u}}{dt^3}(t),\vec{v}\rangle + K_3(t)a(\vec{u}(t_{\bar{n}})\ ;\ \frac{d^2\vec{u}}{dt^2}(t),\vec{v})$$

$$+ K_2(t)a_1(\frac{d^2\vec{u}}{dt^2}(t)\ ;\ \vec{u}(t_{\bar{n}}),\vec{v})\}\,dt$$

$$+ k^2\, a_1\left[\int_{t_n}^{t_{n+2}} K_2(t)\frac{d^2\vec{u}}{dt^2}(t)dt\ ;\ \int_{t_n}^{t_{n+2}} K_3(t)\frac{d^2\vec{u}}{dt^2}(t)dt,\vec{v}\right]\quad \forall\,\vec{v}\in\mathbf{V}.$$

Hence

$$\|\vec{\varepsilon}^{\,n}\|_\star \le c_2k\int_{t_n}^{t_{n+2}}\{\|\frac{d^3\vec{u}}{dt^3}(t)\|_\star + \|\vec{u}(t_{\bar{n}})\|\|\frac{d^2\vec{u}}{dt^2}(t)\|\}dt$$

$$+ c_3k^3\int_{t_n}^{t_{n+2}}\|\frac{d^2\vec{u}}{dt^2}(t)\|^2dt.$$

Therefore

$$\|\vec{\varepsilon}^{n}\|_{*}^{2} \leq c_4 k^3 \int_{t_n}^{t_{n+2}} \{\|\frac{d^3\vec{u}}{dt^3}(t)\|_{*}^{2} + \|\frac{d^2\vec{u}}{dt^2}(t)\|^{2}\}dt ,$$

where the constant $c_4 = c_4\left(\|\vec{u}\|_{\mathcal{C}^o([0,T];\,V)} , \|\frac{d^2u}{dt^2}\|_{L^2(0,T\,;\,V)}\right)$

is independent of n and k. ■

In order to establish the stability of (3.20), we require the following auxiliary lemma :

LEMMA 3.3.

Let (ρ,σ) be the two-step A-stable scheme of example 3.2, and let $\delta = \beta_2 - \frac{\alpha_2}{2} > 0$. Then the coefficients α_i and β_i satisfy the following relation :

$$(3.24)\quad \left\{ \begin{array}{l} 2(\sum_{i=0}^{2} \alpha_i\xi_i)(\sum_{i=0}^{2} \beta_i\xi_i) \geq (\alpha_2^2 + \delta)\xi_2^2 - (2\alpha_2-1)\xi_1^2 - ((\alpha_2-1)^2 + \delta)\xi_o^2 \\ \qquad - 2(\alpha_2(\alpha_2-1) + \delta)(\xi_2\xi_1 - \xi_1\xi_o) , \ \forall\ \xi_o,\xi_1,\xi_2 \in \mathbb{R} . \end{array} \right.$$

PROOF.

By inspection, (3.10) gives immediately :

$$(\sum_{i=0}^{2} \alpha_i\xi_i)(\sum_{i=0}^{2} \beta_i\xi_i) = \{(\alpha_2-1)\xi_o + (1-2\alpha_2)\xi_1 + \alpha_2\xi_2\}\{(\frac{1}{2} - \frac{\alpha_2}{2} + \delta)\xi_o + (\frac{1}{2} - 2\delta)\xi_1 + (\frac{\alpha_2}{2} + \delta)\xi_2\} .$$

The right-hand side can be rearranged as follows :

$$\frac{1}{2}(\alpha_2^2 + \delta)\xi_2^2 - (\alpha_2 - \frac{1}{2})\xi_1^2 - \frac{1}{2}((\alpha_2 - 1)^2 + \delta)\xi_o^2 - (\alpha_2(\alpha_2 - 1) + \delta)(\xi_2\xi_1 - \xi_1\xi_o) + \delta(\alpha_2 - \frac{1}{2})(\xi_2 - 2\,\xi_1 + \xi_o)^2 .$$

Since $\delta > 0$ and $\alpha_2 \geq \frac{1}{2}$, this implies (3.24) . ■

LEMMA 3.4.

Suppose that, in addition to (3.22), $\vec{u} \in \mathcal{C}^0([0,T] ; (L^\infty(\Omega))^n)$. Then the scheme (3.20) satisfies the following stability property :

$$(3.25) \qquad |\vec{e}^m|^2 + k \sum_{n=0}^{m-2} \|\vec{v}^n\|^2 \leq c_1(|\vec{e}^1|^2 + k \sum_{n=0}^{m-2} \|\vec{\varepsilon}^n\|_\star^2)\exp(c_2 t_m) ,$$

for $2 \leq m \leq N$,

where

$$\vec{v}^n = \sum_{i=0}^{2} \beta_i \vec{e}^{n+i}$$

and c_1 , c_2 are constants independent of m and k .

PROOF.

From (3.21) and (3.20), we derive that

$$(3.26) \quad \frac{1}{k} \sum_{i=0}^{2} \alpha_i(\vec{e}^{n+i},\vec{v}) + \sum_{i=0}^{2} \beta_i a(\overline{\vec{u}^n} ; \vec{e}^{n+i},\vec{v}) = < \vec{\varepsilon}^n,\vec{v} >$$

$$- \sum_{i=0}^{2} \beta_i a_1(\sum_{i=0}^{1} \gamma_i \vec{e}^{n+i} ; \vec{u}(t_{n+i}),\vec{v}).$$

Let $\vec{v}^n = \sum_{i=0}^{2} \beta_i \vec{e}^{n+i}$, $\overline{\vec{e}^n} = \sum_{i=0}^{1} \gamma_i \vec{e}^{n+i}$ and take $\vec{v} = \vec{v}^n$ in (3.26) :

$$\frac{1}{k} (\sum_{i=0}^{2} \alpha_i \vec{e}^{n+i}, \sum_{i=0}^{2} \beta_i \vec{e}^{n+i}) + a(\overline{\vec{u}^n} ; \vec{v}^n,\vec{v}^n) = < \vec{\varepsilon}^n , \vec{v}^n >$$

$$- a_1(\overline{\vec{e}^n} ; \sum_{i=0}^{2} \beta_i \vec{u}(t_{n+i}),\vec{v}^n) .$$

Lemma 3.3 implies that

$$2(\sum_{i=0}^{2} \alpha_i \vec{e}^{n+i} , \sum_{i=0}^{2} \beta_i \vec{e}^{n+i}) \geq (\alpha_2^2 + \delta) |\vec{e}^{n+2}|^2 - (2\alpha_2-1)|\vec{e}^{n+1}|^2$$

$$- ((\alpha_2-1)^2 + \delta)|\vec{e}^n|^2 - 2(\alpha_2(\alpha_2-1)+ \delta)\{(\vec{e}^{n+2},\vec{e}^{n+1})-(\vec{e}^{n+1},\vec{e}^n)\} .$$

As a consequence,

$$(a_2^2 + \delta)|\vec{e}^{n+2}|^2 - (2\alpha_2-1)|\vec{e}^{n+1}|^2 - ((\alpha_2-1)^2 + \delta)|\vec{e}^n|^2$$

$$- 2(\alpha_2(\alpha_2-1) + \delta) \{(\vec{e}^{n+2},\vec{e}^{n+1}) - (\vec{e}^{n+1},\vec{e}^n)\} + 2k\nu \|\vec{v}^n\|^2$$

$$\leq 2k \|\vec{\varepsilon}^n\|_\star \|\vec{v}^n\| + 2k|a_1(\overline{\vec{e}^n} ; \vec{v}^n , \sum_{i=0}^{2} \beta_i \vec{u}(t_{n+i}))| .$$

But

$$|a_1(\overline{\vec{e}^n} ; \vec{v}^n, \vec{u}(t_{n+i}))| \leq c_1 \|\vec{u}(t_{n+i})\|_{o,\infty,\Omega} |\overline{\vec{e}^n}| \|\vec{v}^n\| .$$

Therefore, as $\vec{u}(t) \in (L^\infty(\Omega))^n$, we have :

$$2|a_1(\overline{\vec{e}^n} ; \vec{v}^n, \sum_{i=0}^{2} \beta_i \vec{u}(t_{n+i}))| \leq \frac{\nu}{2} \|\vec{v}^n\|^2 + c_2(\nu)|\overline{\vec{e}^n}|^2 .$$

And, as usual, we have :

$$2 \|\vec{\varepsilon}^n\|_\star \|\vec{v}^n\| \leq \frac{\nu}{2} \|\vec{v}^n\|^2 + \frac{2}{\nu} \|\vec{\varepsilon}^n\|_\star^2 .$$

Hence

$$(\alpha_2^2 + \delta)|\vec{e}^{n+2}|^2 - (2\alpha_2 - 1)|\vec{e}^{n+1}|^2 - ((\alpha_2-1)^2 + \delta)|\vec{e}^n|^2$$

$$- 2(\alpha_2(\alpha_2-1)+\delta)\{(\vec{e}^{n+2}, \vec{e}^{n+1}) - (\vec{e}^{n+1}, \vec{e}^n)\} + k\nu \|\vec{v}^n\|^2$$

$$\leq \frac{2k}{\nu} \|\vec{\varepsilon}^n\|_\star^2 + kc_2(\nu)|\overline{\vec{e}^n}|^2 . \tag{3.27}$$

Next, let us sum both sides of (3.27) from $n=0$ to $n = m-2$ and observe that, on one hand :

$$\sum_{n=0}^{m-2} \{(\alpha_2^2 + \delta)|\vec{e}^{n+2}|^2 - (2\alpha_2-1)|\vec{e}^{n+1}|^2 - ((\alpha_2-1)^2 + \delta)|\vec{e}^n|^2\}$$

$$= (\alpha_2^2 + \delta)|\vec{e}^m|^2 + ((\alpha_2-1)^2 + \delta)|\vec{e}^{m-1}|^2 - (\alpha_2^2 + \delta)|\vec{e}^1|^2 ,$$

and on the other hand

$$\sum_{n=0}^{m-2} \{(\vec{e}^{n+2}, \vec{e}^{n+1}) - (\vec{e}^{n+1}, \vec{e}^n)\} = (\vec{e}^m, \vec{e}^{m-1}) .$$

Thus, we obtain

$$(\alpha_2^2+\delta)|\vec{e}^m|^2 + ((\alpha_2-1)^2 + \delta)|\vec{e}^{m-1}|^2 - 2(\alpha_2(\alpha_2-1) + \delta)(\vec{e}^m, \vec{e}^{m-1})$$

$$+ k\nu \sum_{n=0}^{m-2} \|\vec{v}^n\|^2 \leq (\alpha_2^2 + \delta)|\vec{e}^1|^2 + k \sum_{n=0}^{m-2} \{\frac{2}{\nu} \|\vec{\varepsilon}^n\|_\star^2 + c_2(\nu)|\overline{\vec{e}^n}|^2\} . \tag{3.28}$$

But, for all $\varepsilon > 0$, we can write :

$$- 2(\vec{e}^m, \vec{e}^{m-1}) \geq - \varepsilon |\vec{e}^m|^2 - \frac{1}{\varepsilon} |\vec{e}^{m-1}|^2 .$$

Then, by taking $\varepsilon = \dfrac{|\alpha_2(\alpha_2-1)+\delta|}{(\alpha_2-1)^2 + \delta}$, we derive the lower bound

$$- 2|\alpha_2(\alpha_2-1) + \delta|(\vec{e}^m, \vec{e}^{m-1}) \geq - \frac{(\alpha_2(\alpha_{2-1})+\delta)^2}{(\alpha_2-1)^2+\delta} |\vec{e}^m|^2 - ((\alpha_2-1)^2 + \delta)|\vec{e}^{m-1}|^2 .$$

Hence (3.28) yields :

$$(\alpha_2^2 + \delta - \frac{(\alpha_2(\alpha_2-1)+\delta)^2}{(\alpha_2-1)^2+\delta})|\vec{e}^m|^2 + k\nu \sum_{n=0}^{m-2} \|\vec{v}^n\|^2 \leq (\alpha_2^2+\delta)|\vec{e}^1|^2$$

$$+ k \sum_{n=0}^{m-2} \{\frac{2}{\nu}\|\vec{\varepsilon}^n\|_\star^2 + c_2(\nu)|\bar{\vec{e}}^n|^2\} .$$

In the left-hand side, the coefficient of $|\vec{e}^m|^2$ can also be written as :

$$(\alpha_2^2+\delta)\left(1 - \frac{(\alpha_2(\alpha_2-1)+\delta)^2}{(\alpha_2(\alpha_2-1)+\delta)^2+\delta}\right) = \alpha > 0 \quad \text{since} \quad \delta > 0 .$$

Also, in the right-hand side, we can use the upper bound :

$$c_2(\nu) \sum_{n=0}^{m-2} |\bar{\vec{e}}^n|^2 \leq c_3 \sum_{n=0}^{m-1} |\vec{e}^n|^2 .$$

Therefore,

$$\alpha|\vec{e}^m|^2 + k\nu \sum_{n=0}^{m-2} \|\vec{v}^n\|^2 \leq (\alpha_2^2+\delta)|\vec{e}^1|^2 + \frac{2k}{\nu}\sum_{n=0}^{m-2}\|\vec{\varepsilon}^n\|_\star^2 + c_3 k \sum_{n=0}^{m-1} |\vec{e}^n|^2 .$$

Finally, it suffices to apply Lemma 2.4 with $a_m = |\vec{e}^m|^2$, $b_m = \frac{\nu}{\alpha} k \sum_{n=0}^{m-2} \|\vec{v}^n\|^2$,

$b_1 = 0$, $c_m = \frac{1}{\alpha}(\alpha_2^2+\delta)|\vec{e}^1|^2 + \frac{2}{\alpha\nu} k \sum_{n=0}^{m-2} \|\vec{\varepsilon}^n\|_\star^2$, $c_1 = \frac{1}{\alpha}(\alpha_2^2+\delta)|\vec{e}^1|^2$ and $\lambda = \frac{c_3}{\alpha} k$

in order to derive the stability condition (3.25). ∎

By combining Lemmas 3.2 and 3.4, we derive the convergence and error estimate of the solution $\vec{u}^n$ of (3.20).

<u>THEOREM 3.1.</u>

<u>Under the hypotheses of Lemma 3.4 , there exists a constant</u> $c(\vec{u}) > 0$ <u>such that</u>

$$\max_{0 \leq n \leq N} |\vec{u}(t_n) - \vec{u}^n| + (k \sum_{n=0}^{N-2} \| \sum_{i=0}^{2} \beta_i(\vec{u}(t_{n+i}) - \vec{u}^{n+i})\|^2)^{1/2}$$

$$\leq c(\vec{u})\{|\vec{u}(t_1) - \vec{u}_1| + k^2\} .$$

Lecture Notes in Mathematics Vol. 749
ISBN 978-3-540-09557-6

V. Girault and P.-A. Raviart

Finite Element Approximation of the Navier-Stokes Equations

Erratum

p. 21, line 3 of Proof: It should read:

$$(\vec{\varphi}, \vec{\text{curl}}\,\vec{v}) - (\vec{v}, \vec{\text{curl}}\,\vec{\varphi}) = -\int_{\Gamma} (\vec{v} \times \vec{\nu}) \cdot \vec{\varphi}\, d\sigma\,.$$

BIBLIOGRAPHICAL NOTES

Chapter I.

The reader interested in Sobolev spaces can refer to R. Adams [1] or J. Nečas [39]. For more details on elliptic boundary value problems, we refer, for instance, to J.L. Lions & E. Magenes [38] and J. Nečas (loc. cit.).The spaces $H(\mathrm{div}\ ;\ \Omega)$ and $H(\vec{\mathrm{curl}}\ ;\ \Omega)$ are examined by G. Duvaut & J.L. Lions [22] and R. Temam [44]. For complements on the decomposition of vector fields, see O.A. Ladyzhenskaya [34] and R. Temam (loc. cit.).

The crucial result of § 4, Theorem 4.1, is due to F. Brezzi [9]; it is of vital importance in mixed finite element theory. The proof of the regularization algorithm is due to M. Bercovier [5,6] and the duality process of regularization is simply a variant of Uzawa's algorithm (cf. K. Arrow, L. Hurwicz & H. Uzawa [2]).

The theorems asserting the existence and uniqueness of the solution of the Stokes problem are classical. They can be found in such texts like O.A. Ladyzhenskaya (loc. cit.) and R. Temam (loc. cit.). These references also include proofs of the regularity of the solution when the boundary is smooth. In the case of a plane domain with corners, the regularity of the solution is established by V.A. Kondratiev [33] and P. Grisvard [28].

Chapter II.

The analysis of § 1 is essentially due to F. Brezzi [9] with the exception of Theorem 1.2 which is an abstract generalization of the Aubin-Nitsche's trick (cf. J.P. Aubin [3] and J.A. Nitsche [40]). The results of § 1 can be applied to a variety of situations. For instance, they are extensively used by J.M. Thomas [45] and P.A. Raviart & J.M. Thomas [42] in the study of mixed, hybrid equilibrium finite element methods for second order elliptic problems.

The two significant examples analyzed in § 2 are extracted from M. Crouzeix

& P.A. Raviart [18]. This reference contains other examples (all of triangular elements) and more details.

Along the lines of section 2.4, the regularization method and the corresponding penalty method are also discussed by M. Bercovier & M. Engelman [7] and by T.J. Hugues, W.K. Liu & A. Brooks [29] who also consider quadrilateral elements.

Chapter III.

Paragraph 1 reproduces with minor modifications part of the theory contained in F. Brezzi & P.A. Raviart [11] where it is applied to establish the convergence of mixed finite element methods for thin plate problems.

The results of § 2 were originally derived in R. Glowinski [24] and P.G. Ciarlet & P.A. Raviart [15]; they can also be found in P.G. Ciarlet [14]. These convergence results are not optimal : a refined analysis improving the error estimates is given in R. Scholz [43] and V. Girault & P.A. Raviart [23]. For related work, we also refer to R. Glowinski & O. Pironneau [25, 26, 27], C. Johnson & B. Mercier [31] about equilibrium methods and R. Rannacher [41].

Chapter IV.

The approach of § 1 follows the ideas of M. Crouzeix [17].

For complements on the classical theory of the stationary Navier-Stokes equations, we refer to O.A. Ladyzhenskaya [34], J.L. Lions [37] and R. Temam [44].

In § 3, the approximation results for the uniqueness case are due to P. Jamet & P.A. Raviart [30]. In the more general case of a nonsingular solution, the method of proof is that introduced by F. Brezzi [10] for the Von Kármán's equations. The reader can also refer to the results given by H.B. Keller [32] in a more general setting .

Paragraph 4 is a simplified version of V. Girault & P.A. Raviart [23] which deals mainly with nonsingular solutions and analyzes the approximation of the pressure. The reader will find in F. Brezzi, J. Rappaz & P.A. Raviart [12] a generalization of the results of these last two paragraphs to the approximation of branches of nonsingular solutions of nonlinear problems.

Chapter V.

In § 1, the proofs of existence and uniqueness are based on the compactness arguments of J.L. Lions [37]. The results of this paragraph can be found in detail in R. Temam [44].

The convergence of the classical linearized one-step method discussed in § 2 is given by R. Temam (loc. cit.). Here, we complete this result with an error estimate.

In § 3, we adapt a two-step method proposed by M. Zlámal [47] in a slightly different context. This two-step scheme is related to the one-leg multistep method derived by G. Dahlquist [20] for the numerical solution of ordinary differential equations. The tool used here is the energy method based upon the G-stability of the one-leg scheme - a tool implicitly used by M. Zlámal (loc. cit.).

For related results, we refer on one hand to G. Baker [4] who uses discontinuous elements, and on the other hand to M.N. Leroux [36] who studies the convergence of multistep methods for Navier-Stokes equations by operational calculus.

REFERENCES

R.A. Adams,

[1] Sobolev Spaces. Academic Press, New York (1975).

K. Arrow, L. Hurwicz, H. Uzawa,

[2] Studies in Nonlinear Programming. Stanford University Press (1968).

J.P. Aubin,

[3] Approximation of Elliptic Boundary Value Problems. Wiley Interscience, New York (1972).

G.A. Baker,

[4] Simplified proofs of error estimates for the least squares method for Dirichlet's problem. Math. Comp. 27 (1973), pp. 229-235.

M. Bercovier,

[5] Perturbation of mixed variational problems. Application to mixed finite element methods. RAIRO Numer. Anal. 12 n°3 (1978), pp. 211-236.

[6] Thesis, Rouen (1976).

M. Bercovier, M. Engelman,

[7] A finite element for the numerical solution of viscous incompressible flows. J. Comp. Phys. 30 (1979), pp. 181-201.

C. Bernardi,

[8] Thesis (to appear).

F. Brezzi,

[9] On the existence, uniqueness and approximation of saddle point problems arising from Lagrangian multipliers. RAIRO Numer. Anal. 8 - R2 (1974), pp. 129-151.

[10] Finite element approximations of the Von Karmán equations. RAIRO Numer. Anal. 12 n°4 (1978), pp. 303-312.

F. Brezzi, P.A. Raviart,

[11] Mixed finite element methods for fourth order elliptic equations. Topics in Numerical Analysis III. J. Miller (ed.), Academic Press (1977), pp. 33-56.

F. Brezzi, J. Rappaz, P.A. Raviart,

[12] Finite-dimensional approximation of nonlinear problems. Branches of nonsingular solutions. (To appear).

J. Cea,

[13] Optimisation.Théorie et Algorithmes. Dunod, Paris (1971).

P.G. Ciarlet,

[14] The Finite Element Method for Elliptic Problems. North-Holland (1978).

P.G. Ciarlet, P.A. Raviart,

[15] A mixed finite element method for the biharmonic equation. Mathematical Aspects of Finite Elements in Partial Differential Equations. C. de Boor (ed.), Academic Press (1974), pp. 125-145.

E.A. Coddington, N. Levinson,

[16] Theory of Ordinary Differential Equations. Mc Graw-Hill (1955).

M. Crouzeix

[17] Etude d'une méthode de linéarisation. Résolution numérique des équations de Stokes stationnaires. Application aux équations de Navier-Stokes stationnaires. IRIA, cahier n°12 (1974), pp. 141-244.

M. Crouzeix, P.A. Raviart,

[18] Conforming and non conforming finite element methods for solving the stationary Stokes equations. RAIRO Numer. Anal. 7-R3 (1973), pp. 33-76.

G. Dahlquist,

[19] Convergence and stability in the numerical integration of ordinary differential equations. Math. Scand. 4 (1956), pp. 33-53.

[20] On the relation of G-Stability to other stability concepts for linear multistep methods. Topics in Numerical Analysis III. J. Miller (ed.), Academic Press (1977), pp. 67-80.

J. Jr. Douglas, T. Dupont,

[21] Galerkin methods for parabolic equations. SIAM J. Numer. Anal. 11 (1974), pp. 392-410.

G. Duvaut, J.L. Lions,

[22] Les Inéquations en Mécanique et en Physique. Dunod, Paris (1971).

V. Girault, P.A. Raviart,

[23] An analysis of a mixed finite element method for the Navier-Stokes equations. Numer. Math. (to appear).

R. Glowinski,

[24] Approximations externes, par éléments finis de Lagrange d'ordre un et deux, du problème de Dirichlet pour l'opérateur biharmonique. Méthode itérative de résolution des problèmes approchés. Topics in Numerical Analysis. J. Miller (ed.), Academic Press (1973), pp. 123-171.

R. Glowinski, O. Pironneau,

[25] Approximation par éléments finis mixtes du problème de Stokes en formulation vitesse-pression. Convergence des solutions approchées. C.R. Acad. Sci. Paris, Série A , 286 (1978), pp.181-183.

[26] Approximation par éléments finis mixtes du problème de Stokes en formulation vitesse-pression. Résolution des problèmes approchés. Ibid 286 (1978), pp. 225-228.

[27] Numerical methods for the first biharmonic equation and the two-dimensional Stokes problem. SIAM Review 21 n°2 (1979), pp.167-212.

P. Grisvard,

[28] Singularité des solutions du problème de Stokes dans un polygone. Séminaires d'Analyse Numérique, Paris (1978).

T.J. Hughes, W.K. Liu, A. Brooks,

[29] Finite element analysis of incompressible viscous flows by the penalty function formulation. J. Comp. Phys. 30 (1979), pp. 1-60.

P. Jamet, P.A. Raviart,

[30] Numerical Solution of the Stationary Navier-Stokes Equations by Finite Element Methods. Lecture Notes in Computer Science, Springer-Verlag 10 (1973), pp. 193-223.

C. Johnson, B. Mercier,

[31] Some equilibrium finite element methods for two-dimensional problems in continuum mechanics. Energy Methods in Finite Element Analysis. R. Glowinski, E. Y. Rodin, O.C. Zienkiewicz (ed.), Wiley, Chichester (1979), pp. 213-224.

H.B. Keller,

[32] Approximation methods for nonlinear problems with applications to two-points boundary value problems. Math. Comp. 29 (1975), pp. 464-474.

V.A. Kondrat'ev,

[33] Boundary problems for elliptic equations in domains with conical or angular points. Trans. Moscow Math. Soc. (1967), pp. 227-313.

O.A. Ladyzhenskaya

[34] The Mathematical Theory of Viscous Incompressible Flow. Gordon and Breach, New York (1969).

P.D. Lax, A.N. Milgram

[35] Parabolic Equations. Contributions to the Theory of Partial Differential Equations, Princeton (1954).

M. N. Leroux,

[36] Thesis (to appear).

J. L. Lions,

[37] Quelques Méthodes de Résolution des Problèmes aux Limites non Linéaires. Dunod, Paris (1969).

J.L. Lions, E. Magenes,

[38] Nonhomogeneous Boundary Value Problems and Applications Springer-Verlag, Berlin (1972).

J. Nečas,

[39] Les Méthodes Directes en Théorie des Equations Elliptiques. Masson, Paris (1967).

J.A. Nitsche

[40] Ein kriterium für die quasi-optimalitat des Ritzchen Verfahrens. Numer. Math. 11 (1968), pp. 346-348.

R. Rannacher

[41] Punktweise Konvergenz der Methode der Finiten Elemente beim Plattenproblem. Manuscripta Math. 19 (1976), pp. 401-416.

P.A. Raviart, J.M. Thomas,

[42] A mixed finite element method for second order elliptic problems. Lecture Notes in Mathematics, Springer-Verlag, 606, pp. 292-315.

R. Scholz,

[43] A mixed method for 4th order problems using linear finite elements. RAIRO Numer. Anal. 12 n°1 (1978), pp. 85-90.

R. Temam,

[44] Navier-Stokes Equations. North-Holland, Amsterdam (1977).

J.M. Thomas,

[45] Sur l'analyse numérique des méthodes d'éléments finis hybrides et mixtes. Thesis, Paris (1977).

K. Yosida,

[46] Functional Analysis. Springer-Verlag, Berlin (1965).

M. Zlámal

[47] Finite element methods for non linear parabolic equations. RAIRO Numer. Anal. 11 n°1 (1977), pp. 93-107.

I N D E X

APPENDIX

THEOREM 2.1.

The space $[\mathcal{D}(\bar{\Omega})]^n$ is dense in $H(\text{div}\ ;\ \Omega)$.

PROOF.

1°) Let ℓ belong to $(H(\text{div}\ ;\ \Omega))'$, the dual space of $H(\text{div}\ ;\ \Omega)$ for the norm (2.1). As $H(\text{div}\ ;\ \Omega)$ is a Hilbert space, there exists according to the F. Riesz representation theorem an element $\vec{f}$ of $H(\text{div}\ ;\ \Omega)$ such that

$$\ell(\vec{q}) = (\vec{f},\vec{q}) + (f_{n+1},\text{div}\ \vec{q}) \quad \forall\ \vec{q} \in H(\text{div}\ ;\ \Omega) \tag{2.2}$$

where

$$f_{n+1} = \text{div}\ \vec{f}\ .$$

2°) Now, assume that ℓ is any element of $(H(\text{div}\ ;\ \Omega))'$ that satisfies :

$$\ell(\vec{u}) = 0 \quad \forall\ \vec{u} \in [\mathcal{D}(\bar{\Omega})]^n\ . \tag{2.3}$$

If we can show that $\ell \equiv 0$ then, as a consequence of the Hahn-Banach theorem, this will imply that $[\mathcal{D}(\bar{\Omega})]^n$ is dense in $H(\text{div}\ ;\ \Omega)$.

Let $\vec{f} \in H(\text{div}\ ;\ \Omega)$ be associated with ℓ by (2.2). Thus, (2.3) states that

$$(\vec{f},\vec{u}) + (f_{n+1},\text{div}\ \vec{u}) = 0 \quad \forall\ \vec{u} \in [\mathcal{D}(\bar{\Omega})]^n\ .$$

This can also be written as :

$$\int_{\mathbb{R}^n} \{\sum_{i=1}^{n} \tilde{f}_i u_i + \tilde{f}_{n+1}\ \text{div}\ \vec{u}\} = 0 \quad \forall\ \vec{u} \in [\mathcal{D}(\mathbb{R}^n)]^n\ ,$$

where the tilde denotes the extension of the function by zero outside Ω. Hence

$$< \overrightarrow{\text{grad}}\ \tilde{f}_{n+1},\vec{u} > = \int_{\mathbb{R}^n} \sum_{i=1}^{n} \tilde{f}_i u_i\ dx \quad \forall\ \vec{u} \in [\mathcal{D}(\mathbb{R}^n)]^n\ ,$$

and therefore

$$\tilde{f}_i = \frac{\partial \tilde{f}_{n+1}}{\partial x_i} \quad \text{in} \quad \mathcal{D}'(\mathbb{R}^n)\ ,\ 1 \leqslant i \leqslant n\ .$$

As $\tilde{f}_i \in L^2(\mathbb{R}^n)$, for $1 \leqslant i \leqslant n+1$, this means that $\tilde{f}_{n+1} \in H^1(\mathbb{R}^n)$.
In turn, this implies that $f_{n+1} \in H_0^1(\Omega)$ (this is a particular consequence of the density of $\mathcal{D}(\bar{\Omega})$ in $H^1(\Omega)$). Moreover,

$$\vec{f} = \text{grad } f_{n+1} \quad \text{in} \quad \Omega .$$

From this last result, the density of $\mathcal{D}(\Omega)$ in $H_o^1(\Omega)$, formula (2.2) and the hypothesis (2.3) we readily derive that

$$\ell(\vec{q}) = (\text{grad } f_{n+1}, \vec{q}) + (f_{n+1}, \text{div } \vec{q}) = 0 \quad \forall \vec{q} \in H(\text{div } ; \Omega) ,$$

thus proving that $\ell = 0$. ■

THEOREM 2.3.

We have

$$\text{Ker } \gamma_\nu = H_o(\text{div } ; \Omega) .$$

Proof.

The idea of the proof is similar to that of Theorem 2.1. We shall show that each element ℓ of $(\text{Ker } \gamma_\nu)'$ that vanishes on $(\mathcal{D}(\Omega))^n$ also vanishes on Ker γ_ν ; therefore by the Hahn-Banach theorem, this will imply that $(\mathcal{D}(\Omega))^n$ is dense in Ker γ_ν .

Again, let $\vec{f}$ be the element of Ker γ_ν associated with ℓ by the F. Riesz representation theorem :

$$\ell(\vec{q}) = (\vec{f}, \vec{q}) + (f_{n+1}, \text{div } \vec{q}) \quad \forall \vec{q} \in \text{Ker } \gamma_\nu ,$$

where

$$f_{n+1} = \text{div } \vec{f} .$$

Therefore, by hypothesis :

$$(\vec{f}, \vec{u}) + (f_{n+1}, \text{div } \vec{u}) = 0 \quad \forall \vec{u} \in (\mathcal{D}(\Omega))^n ;$$

hence

$$\text{grad } f_{n+1} = \vec{f} \quad \text{in} \quad (\mathcal{D}'(\Omega))^n .$$

Thus, $f_{n+1} \in H^1(\Omega)$ and Green's formula (2.7) yields :

$$\ell(\vec{q}) = < \gamma_\nu \vec{q}, \gamma_o f_{n+1} >_\Gamma = 0 \quad \forall \vec{q} \in \text{Ker } \gamma_\nu \quad . \quad ■$$

Lecture Notes in Mathematics

For information about Vols. 1–521, please contact your bookseller or Springer-Verlag.

Vol. 522: C. O. Bloom and N. D. Kazarinoff, Short Wave Radiation Problems in Inhomogeneous Media: Asymptotic Solutions. V. 104 pages. 1976.

Vol. 523: S. A. Albeverio and R. J. Høegh-Krohn, Mathematical Theory of Feynman Path Integrals. IV, 139 pages. 1976.

Vol. 524: Séminaire Pierre Lelong (Analyse) Année 1974/75. Edité par P. Lelong. V, 222 pages. 1976.

Vol. 525: Structural Stability, the Theory of Catastrophes, and Applications in the Sciences. Proceedings 1975. Edited by P. Hilton. VI, 408 pages. 1976.

Vol. 526: Probability in Banach Spaces. Proceedings 1975. Edited by A. Beck. VI, 290 pages. 1976.

Vol. 527: M. Denker, Ch. Grillenberger, and K. Sigmund, Ergodic Theory on Compact Spaces. IV, 360 pages. 1976.

Vol. 528: J. E. Humphreys, Ordinary and Modular Representations of Chevalley Groups. III, 127 pages. 1976.

Vol. 529: J. Grandell, Doubly Stochastic Poisson Processes. X, 234 pages. 1976.

Vol. 530: S. S. Gelbart, Weil's Representation and the Spectrum of the Metaplectic Group. VII, 140 pages. 1976.

Vol. 531: Y.-C. Wong, The Topology of Uniform Convergence on Order-Bounded Sets. VI, 163 pages. 1976.

Vol. 532: Théorie Ergodique. Proceedings 1973/1974. Edité par J.-P. Conze and M. S. Keane. VIII, 227 pages. 1976.

Vol. 533: F. R. Cohen, T. J. Lada, and J. P. May, The Homology of Iterated Loop Spaces. IX, 490 pages. 1976.

Vol. 534: C. Preston, Random Fields. V, 200 pages. 1976.

Vol. 535: Singularités d'Applications Differentiables. Plans-sur-Bex. 1975. Edité par O. Burlet et F. Ronga. V, 253 pages. 1976.

Vol. 536: W. M. Schmidt, Equations over Finite Fields. An Elementary Approach. IX, 267 pages. 1976.

Vol. 537: Set Theory and Hierarchy Theory. Bierutowice, Poland 1975. A Memorial Tribute to Andrzej Mostowski. Edited by W. Marek, M. Srebrny and A. Zarach. XIII, 345 pages. 1976.

Vol. 538: G. Fischer, Complex Analytic Geometry. VII, 201 pages. 1976.

Vol. 539: A. Badrikian, J. F. C. Kingman et J. Kuelbs, Ecole d'Eté de Probabilités de Saint Flour V-1975. Edité par P.-L. Hennequin. IX, 314 pages. 1976.

Vol. 540: Categorical Topology, Proceedings 1975. Edited by E. Binz and H. Herrlich. XV, 719 pages. 1976.

Vol. 541: Measure Theory, Oberwolfach 1975. Proceedings. Edited by A. Bellow and D. Kölzow. XIV, 430 pages. 1976.

Vol. 542: D. A. Edwards and H. M. Hastings, Čech and Steenrod Homotopy Theories with Applications to Geometric Topology. VII, 296 pages. 1976.

Vol. 543: Nonlinear Operators and the Calculus of Variations, Bruxelles 1975. Edited by J. P. Gossez, E. J. Lami Dozo, J. Mawhin, and L. Waelbroeck, VII, 237 pages. 1976.

Vol. 544: Robert P. Langlands, On the Functional Equations Satisfied by Eisenstein Series. VII, 337 pages. 1976.

Vol. 545: Noncommutative Ring Theory. Kent State 1975. Edited by J. H. Cozzens and F. L. Sandomierski. V, 212 pages. 1976.

Vol. 546: K. Mahler, Lectures on Transcendental Numbers. Edited and Completed by B. Diviš and W. J. Le Veque. XXI, 254 pages. 1976.

Vol. 547: A. Mukherjea and N. A. Tserpes, Measures on Topological Semigroups: Convolution Products and Random Walks. V, 197 pages. 1976.

Vol. 548: D. A. Hejhal, The Selberg Trace Formula for PSL (2,ℝ). Volume I. VI, 516 pages. 1976.

Vol. 549: Brauer Groups, Evanston 1975. Proceedings. Edited by D. Zelinsky. V, 187 pages. 1976.

Vol. 550: Proceedings of the Third Japan – USSR Symposium on Probability Theory. Edited by G. Maruyama and J. V. Prokhorov. VI, 722 pages. 1976.

Vol. 551: Algebraic K-Theory, Evanston 1976. Proceedings. Edited by M. R. Stein. XI, 409 pages. 1976.

Vol. 552: C. G. Gibson, K. Wirthmüller, A. A. du Plessis and E. J. N. Looijenga. Topological Stability of Smooth Mappings. V, 155 pages. 1976.

Vol. 553: M. Petrich, Categories of Algebraic Systems. Vector and Projective Spaces, Semigroups, Rings and Lattices. VIII, 217 pages. 1976.

Vol. 554: J. D. H. Smith, Mal'cev Varieties. VIII, 158 pages. 1976.

Vol. 555: M. Ishida, The Genus Fields of Algebraic Number Fields. VII, 116 pages. 1976.

Vol. 556: Approximation Theory. Bonn 1976. Proceedings. Edited by R. Schaback and K. Scherer. VII, 466 pages. 1976.

Vol. 557: W. Iberkleid and T. Petrie, Smooth S^1 Manifolds. III, 163 pages. 1976.

Vol. 558: B. Weisfeiler, On Construction and Identification of Graphs. XIV, 237 pages. 1976.

Vol. 559: J.-P. Caubet, Le Mouvement Brownien Relativiste. IX 212 pages. 1976.

Vol. 560: Combinatorial Mathematics, IV, Proceedings 1975. Edite by L. R. A. Casse and W. D. Wallis. VII, 249 pages. 1976.

Vol. 561: Function Theoretic Methods for Partial Differential Equation Darmstadt 1976. Proceedings. Edited by V. E. Meister, N. Wec and W. L. Wendland. XVIII, 520 pages. 1976.

Vol. 562: R. W. Goodman, Nilpotent Lie Groups: Structure an Applications to Analysis. X, 210 pages. 1976.

Vol. 563: Séminaire de Théorie du Potentiel. Paris, No. 2. Proceeding 1975–1976. Edited by F. Hirsch and G. Mokobodzki. VI, 292 page 1976.

Vol. 564: Ordinary and Partial Differential Equations, Dundee 197 Proceedings. Edited by W. N. Everitt and B. D. Sleeman. XVIII, 5 pages. 1976.

Vol. 565: Turbulence and Navier Stokes Equations. Proceeding 1975. Edited by R. Temam. IX, 194 pages. 1976.

Vol. 566: Empirical Distributions and Processes. Oberwolfach 197 Proceedings. Edited by P. Gaenssler and P. Révész. VII, 146 page 1976.

Vol. 567: Séminaire Bourbaki vol. 1975/76. Exposés 471–488. 303 pages. 1977.

Vol. 568: R. E. Gaines and J. L. Mawhin, Coincidence Degree, a Nonlinear Differential Equations. V, 262 pages. 1977.

Vol. 569: Cohomologie Etale SGA 4½. Séminaire de Géomét Algébrique du Bois-Marie. Edité par P. Deligne. V, 312 pages. 197

Vol. 570: Differential Geometrical Methods in Mathematical Physi Bonn 1975. Proceedings. Edited by K. Bleuler and A. Reetz. V 576 pages. 1977.

Vol. 571: Constructive Theory of Functions of Several Variabl Oberwolfach 1976. Proceedings. Edited by W. Schempp and K. Z ler. VI. 290 pages. 1977

Vol. 572: Sparse Matrix Techniques, Copenhagen 1976. Edited V. A. Barker. V, 184 pages. 1977.

Vol. 573: Group Theory, Canberra 1975. Proceedings. Edited R. A. Bryce, J. Cossey and M. F. Newman. VII, 146 pages. 197

Vol. 574: J. Moldestad, Computations in Higher Types. IV, 2 pages. 1977.

Vol. 575: K-Theory and Operator Algebras, Athens, Georgia 19 Edited by B. B. Morrel and I. M. Singer. VI, 191 pages. 1977.

Vol. 576: V. S. Varadarajan, Harmonic Analysis on Real Reduc Groups. VI, 521 pages. 1977.

Vol. 577: J. P. May, E_∞ Ring Spaces and E_∞ Ring Spectra. 268 pages. 1977.

Vol. 578: Séminaire Pierre Lelong (Analyse) Année 1975/76. E par P. Lelong. VI, 327 pages. 1977.

Vol. 579: Combinatoire et Représentation du Groupe Symétri Strasbourg 1976. Proceedings 1976. Edité par D. Foata. IV, pages. 1977.

Vol. 580: C. Castaing and M. Valadier, Convex Analysis and Measurable Multifunctions. VIII, 278 pages. 1977.

Vol. 581: Séminaire de Probabilités XI, Université de Strasbourg. Proceedings 1975/1976. Edité par C. Dellacherie, P. A. Meyer et M. Weil. VI, 574 pages. 1977.

Vol. 582: J. M. G. Fell, Induced Representations and Banach *-Algebraic Bundles. IV, 349 pages. 1977.

Vol. 583: W. Hirsch, C. C. Pugh and M. Shub, Invariant Manifolds. IV, 149 pages. 1977.

Vol. 584: C. Brezinski, Accélération de la Convergence en Analyse Numérique. IV, 313 pages. 1977.

Vol. 585: T. A. Springer, Invariant Theory. VI, 112 pages. 1977.

Vol. 586: Séminaire d'Algèbre Paul Dubreil, Paris 1975–1976 (29ème Année). Edited by M. P. Malliavin. VI, 188 pages. 1977.

Vol. 587: Non-Commutative Harmonic Analysis. Proceedings 1976. Edited by J. Carmona and M. Vergne. IV, 240 pages. 1977.

Vol. 588: P. Molino, Théorie des G-Structures: Le Problème d'Equivalence. VI, 163 pages. 1977.

Vol. 589: Cohomologie l-adique et Fonctions L. Séminaire de Géométrie Algébrique du Bois-Marie 1965–66, SGA 5. Edité par L. Illusie. XII, 484 pages. 1977.

Vol. 590: H. Matsumoto, Analyse Harmonique dans les Systèmes de Tits Bornologiques de Type Affine. IV, 219 pages. 1977.

Vol. 591: G. A. Anderson, Surgery with Coefficients. VIII, 157 pages. 1977.

Vol. 592: D. Voigt, Induzierte Darstellungen in der Theorie der endlichen, algebraischen Gruppen. V, 413 Seiten. 1977.

Vol. 593: K. Barbey and H. König, Abstract Analytic Function Theory and Hardy Algebras. VIII, 260 pages. 1977.

Vol. 594: Singular Perturbations and Boundary Layer Theory, Lyon 1976. Edited by C. M. Brauner, B. Gay, and J. Mathieu. VIII, 539 pages. 1977.

Vol. 595: W. Hazod, Stetige Faltungshalbgruppen von Wahrscheinlichkeitsmaßen und erzeugende Distributionen. XIII, 157 Seiten. 1977.

Vol. 596: K. Deimling, Ordinary Differential Equations in Banach Spaces. VI, 137 pages. 1977.

Vol. 597: Geometry and Topology, Rio de Janeiro, July 1976. Proceedings. Edited by J. Palis and M. do Carmo. VI, 866 pages. 1977.

Vol. 598: J. Hoffmann-Jørgensen, T. M. Liggett et J. Neveu, Ecole d'Eté de Probabilités de Saint-Flour VI – 1976. Edité par P.-L. Hennequin. XII, 447 pages. 1977.

Vol. 599: Complex Analysis, Kentucky 1976. Proceedings. Edited by J. D. Buckholtz and T. J. Suffridge. X, 159 pages. 1977.

Vol. 600: W. Stoll, Value Distribution on Parabolic Spaces. VIII, 216 pages. 1977.

Vol. 601: Modular Functions of one Variable V, Bonn 1976. Proceedings. Edited by J.-P. Serre and D. B. Zagier. VI, 294 pages. 1977.

Vol. 602: J. P. Brezin, Harmonic Analysis on Compact Solvmanifolds. VIII, 179 pages. 1977.

Vol. 603: B. Moishezon, Complex Surfaces and Connected Sums of Complex Projective Planes. IV, 234 pages. 1977.

Vol. 604: Banach Spaces of Analytic Functions, Kent, Ohio 1976. Proceedings. Edited by J. Baker, C. Cleaver and Joseph Diestel. VI, 141 pages. 1977.

Vol. 605: Sario et al., Classification Theory of Riemannian Manifolds. XX, 498 pages. 1977.

Vol. 606: Mathematical Aspects of Finite Element Methods. Proceedings 1975. Edited by I. Galligani and E. Magenes. VI, 362 pages. 1977.

Vol. 607: M. Métivier, Reelle und Vektorwertige Quasimartingale und die Theorie der Stochastischen Integration. X, 310 Seiten. 1977.

Vol. 608: Bigard et al., Groupes et Anneaux Réticulés. XIV, 334 pages. 1977.

Vol. 609: General Topology and Its Relations to Modern Analysis and Algebra IV. Proceedings 1976. Edited by J. Novák. XVIII, 225 pages. 1977.

Vol. 610: G. Jensen, Higher Order Contact of Submanifolds of Homogeneous Spaces. XII, 154 pages. 1977.

Vol. 611: M. Makkai and G. E. Reyes, First Order Categorical Logic. VIII, 301 pages. 1977.

Vol. 612: E. M. Kleinberg, Infinitary Combinatorics and the Axiom of Determinateness. VIII, 150 pages. 1977.

Vol. 613: E. Behrends et al., L^p-Structure in Real Banach Spaces. X, 108 pages. 1977.

Vol. 614: H. Yanagihara, Theory of Hopf Algebras Attached to Group Schemes. VIII, 308 pages. 1977.

Vol. 615: Turbulence Seminar, Proceedings 1976/77. Edited by P. Bernard and T. Ratiu. VI, 155 pages. 1977.

Vol. 616: Abelian Group Theory, 2nd New Mexico State University Conference, 1976. Proceedings. Edited by D. Arnold, R. Hunter and E. Walker. X, 423 pages. 1977.

Vol. 617: K. J. Devlin, The Axiom of Constructibility: A Guide for the Mathematician. VIII, 96 pages. 1977.

Vol. 618: I. I. Hirschman, Jr. and D. E. Hughes, Extreme Eigen Values of Toeplitz Operators. VI, 145 pages. 1977.

Vol. 619: Set Theory and Hierarchy Theory V, Bierutowice 1976. Edited by A. Lachlan, M. Srebrny, and A. Zarach. VIII, 358 pages. 1977.

Vol. 620: H. Popp, Moduli Theory and Classification Theory of Algebraic Varieties. VIII, 189 pages. 1977.

Vol. 621: Kauffman et al., The Deficiency Index Problem. VI, 112 pages. 1977.

Vol. 622: Combinatorial Mathematics V, Melbourne 1976. Proceedings. Edited by C. Little. VIII, 213 pages. 1977.

Vol. 623: I. Erdelyi and R. Lange, Spectral Decompositions on Banach Spaces. VIII, 122 pages. 1977.

Vol. 624: Y. Guivarc'h et al., Marches Aléatoires sur les Groupes de Lie. VIII, 292 pages. 1977.

Vol. 625: J. P. Alexander et al., Odd Order Group Actions and Witt Classification of Innerproducts. IV, 202 pages. 1977.

Vol. 626: Number Theory Day, New York 1976. Proceedings. Edited by M. B. Nathanson. VI, 241 pages. 1977.

Vol. 627: Modular Functions of One Variable VI, Bonn 1976. Proceedings. Edited by J.-P. Serre and D. B. Zagier. VI, 339 pages. 1977.

Vol. 628: H. J. Baues, Obstruction Theory on the Homotopy Classification of Maps. XII, 387 pages. 1977.

Vol. 629: W. A. Coppel, Dichotomies in Stability Theory. VI, 98 pages. 1978.

Vol. 630: Numerical Analysis, Proceedings, Biennial Conference, Dundee 1977. Edited by G. A. Watson. XII, 199 pages. 1978.

Vol. 631: Numerical Treatment of Differential Equations. Proceedings 1976. Edited by R. Bulirsch, R. D. Grigorieff, and J. Schröder. X, 219 pages. 1978.

Vol. 632: J.-F. Boutot, Schéma de Picard Local. X, 165 pages. 1978.

Vol. 633: N. R. Coleff and M. E. Herrera, Les Courants Résiduels Associés à une Forme Méromorphe. X, 211 pages. 1978.

Vol. 634: H. Kurke et al., Die Approximationseigenschaft lokaler Ringe. IV, 204 Seiten. 1978.

Vol. 635: T. Y. Lam, Serre's Conjecture. XVI, 227 pages. 1978.

Vol. 636: Journées de Statistique des Processus Stochastiques, Grenoble 1977, Proceedings. Edité par Didier Dacunha-Castelle et Bernard Van Cutsem. VII, 202 pages. 1978.

Vol. 637: W. B. Jurkat, Meromorphe Differentialgleichungen. VII, 194 Seiten. 1978.

Vol. 638: P. Shanahan, The Atiyah-Singer Index Theorem, An Introduction. V, 224 pages. 1978.

Vol. 639: N. Adasch et al., Topological Vector Spaces. V, 125 pages. 1978.

Vol. 640: J. L. Dupont, Curvature and Characteristic Classes. X, 175 pages. 1978.

Vol. 641: Séminaire d'Algèbre Paul Dubreil, Proceedings Paris 1976–1977. Edité par M. P. Malliavin. IV, 367 pages. 1978.

Vol. 642: Theory and Applications of Graphs, Proceedings, Michigan 1976. Edited by Y. Alavi and D. R. Lick. XIV, 635 pages. 1978.

Vol. 643: M. Davis, Multiaxial Actions on Manifolds. VI, 141 pages. 1978.

Vol. 644: Vector Space Measures and Applications I, Proceedings 1977. Edited by R. M. Aron and S. Dineen. VIII, 451 pages. 1978.

Vol. 645: Vector Space Measures and Applications II, Proceedings 1977. Edited by R. M. Aron and S. Dineen. VIII, 218 pages. 1978.

Vol. 646: O. Tammi, Extremum Problems for Bounded Univalent Functions. VIII, 313 pages. 1978.

Vol. 647: L. J. Ratliff, Jr., Chain Conjectures in Ring Theory. VIII, 133 pages. 1978.

Vol. 648: Nonlinear Partial Differential Equations and Applications, Proceedings, Indiana 1976–1977. Edited by J. M. Chadam. VI, 206 pages. 1978.

Vol. 649: Séminaire de Probabilités XII, Proceedings, Strasbourg, 1976–1977. Edité par C. Dellacherie, P. A. Meyer et M. Weil. VIII, 805 pages. 1978.

Vol. 650: C*-Algebras and Applications to Physics. Proceedings 1977. Edited by H. Araki and R. V. Kadison. V, 192 pages. 1978.

Vol. 651: P. W. Michor, Functors and Categories of Banach Spaces. VI, 99 pages. 1978.

Vol. 652: Differential Topology, Foliations and Gelfand-Fuks-Cohomology, Proceedings 1976. Edited by P. A. Schweitzer. XIV, 252 pages. 1978.

Vol. 653: Locally Interacting Systems and Their Application in Biology. Proceedings, 1976. Edited by R. L. Dobrushin, V. I. Kryukov and A. L. Toom. XI, 202 pages. 1978.

Vol. 654: J. P. Buhler, Icosahedral Golois Representations. III, 143 pages. 1978.

Vol. 655: R. Baeza, Quadratic Forms Over Semilocal Rings. VI, 199 pages. 1978.

Vol. 656: Probability Theory on Vector Spaces. Proceedings, 1977. Edited by A. Weron. VIII, 274 pages. 1978.

Vol. 657: Geometric Applications of Homotopy Theory I, Proceedings 1977. Edited by M. G. Barratt and M. E. Mahowald. VIII, 459 pages. 1978.

Vol. 658: Geometric Applications of Homotopy Theory II, Proceedings 1977. Edited by M. G. Barratt and M. E. Mahowald. VIII, 487 pages. 1978.

Vol. 659: Bruckner, Differentiation of Real Functions. X, 247 pages. 1978.

Vol. 660: Equations aux Dérivée Partielles. Proceedings, 1977. Edité par Pham The Lai. VI, 216 pages. 1978.

Vol. 661: P. T. Johnstone, R. Paré, R. D. Rosebrugh, D. Schumacher, R. J. Wood, and G. C. Wraith, Indexed Categories and Their Applications. VII, 260 pages. 1978.

Vol. 662: Akin, The Metric Theory of Banach Manifolds. XIX, 306 pages. 1978.

Vol. 663: J. F. Berglund, H. D. Junghenn, P. Milnes, Compact Right Topological Semigroups and Generalizations of Almost Periodicity. X, 243 pages. 1978.

Vol. 664: Algebraic and Geometric Topology, Proceedings, 1977. Edited by K. C. Millett. XI, 240 pages. 1978.

Vol. 665: Journées d'Analyse Non Linéaire. Proceedings, 1977. Edité par P. Bénilan et J. Robert. VIII, 256 pages. 1978.

Vol. 666: B. Beauzamy, Espaces d'Interpolation Réels: Topologie et Géometrie. X, 104 pages. 1978.

Vol. 667: J. Gilewicz, Approximants de Padé. XIV, 511 pages. 1978.

Vol. 668: The Structure of Attractors in Dynamical Systems. Proceedings, 1977. Edited by J. C. Martin, N. G. Markley and W. Perrizo. VI, 264 pages. 1978.

Vol. 669: Higher Set Theory. Proceedings, 1977. Edited by G. H. Müller and D. S. Scott. XII, 476 pages. 1978.

Vol. 670: Fonctions de Plusieurs Variables Complexes III, Proceedings, 1977. Edité par F. Norguet. XII, 394 pages. 1978.

Vol. 671: R. T. Smythe and J. C. Wierman, First-Passage Perculation on the Square Lattice. VIII, 196 pages. 1978.

Vol. 672: R. L. Taylor, Stochastic Convergence of Weighted Sums of Random Elements in Linear Spaces. VII, 216 pages. 1978.

Vol. 673: Algebraic Topology, Proceedings 1977. Edited by P. Hoffman, R. Piccinini and D. Sjerve. VI, 278 pages. 1978.

Vol. 674: Z. Fiedorowicz and S. Priddy, Homology of Classica Groups Over Finite Fields and Their Associated Infinite Loop Spaces VI, 434 pages. 1978.

Vol. 675: J. Galambos and S. Kotz, Characterizations of Probability Distributions. VIII, 169 pages. 1978.

Vol. 676: Differential Geometrical Methods in Mathematical Physics II, Proceedings, 1977. Edited by K. Bleuler, H. R. Petry and A. Reetz VI, 626 pages. 1978.

Vol. 677: Séminaire Bourbaki, vol. 1976/77, Exposés 489–506. IV 264 pages. 1978.

Vol. 678: D. Dacunha-Castelle, H. Heyer et B. Roynette. Ecole d'Eté de Probabilités de Saint-Flour. VII-1977. Edité par P. L. Hennequin IX, 379 pages. 1978.

Vol. 679: Numerical Treatment of Differential Equations in Applica tions, Proceedings, 1977. Edited by R. Ansorge and W. Törnig. IX 163 pages. 1978.

Vol. 680: Mathematical Control Theory, Proceedings, 1977. Edite by W. A. Coppel. IX, 257 pages. 1978.

Vol. 681: Séminaire de Théorie du Potentiel Paris, No. 3, Directeurs M. Brelot, G. Choquet et J. Deny. Rédacteurs: F. Hirsch et G. Moko bodzki. VII, 294 pages. 1978.

Vol. 682: G. D. James, The Representation Theory of the Symmetri Groups. V, 156 pages. 1978.

Vol. 683: Variétés Analytiques Compactes, Proceedings, 197 Edité par Y. Hervier et A. Hirschowitz. V, 248 pages. 1978.

Vol. 684: E. E. Rosinger, Distributions and Nonlinear Partial Di ferential Equations. XI, 146 pages. 1978.

Vol. 685: Knot Theory, Proceedings, 1977. Edited by J. C. Hausman VII, 311 pages. 1978.

Vol. 686: Combinatorial Mathematics, Proceedings, 1977. Edite by D. A. Holton and J. Seberry. IX, 353 pages. 1978.

Vol. 687: Algebraic Geometry, Proceedings, 1977. Edited by L. Olson. V, 244 pages. 1978.

Vol. 688: J. Dydak and J. Segal, Shape Theory. VI, 150 pages. 197

Vol. 689: Cabal Seminar 76–77, Proceedings, 1976–77. Edited A.S. Kechris and Y. N. Moschovakis. V, 282 pages. 1978.

Vol. 690: W. J. J. Rey, Robust Statistical Methods. VI, 128 pages. 197

Vol. 691: G. Viennot, Algèbres de Lie Libres et Monoïdes Libre III, 124 pages. 1978.

Vol. 692: T. Husain and S. M. Khaleelulla, Barrelledness in Top logical and Ordered Vector Spaces. IX, 258 pages. 1978.

Vol. 693: Hilbert Space Operators, Proceedings, 1977. Edited J. M. Bachar Jr. and D. W. Hadwin. VIII, 184 pages. 1978.

Vol. 694: Séminaire Pierre Lelong – Henri Skoda (Analyse) Ann 1976/77. VII, 334 pages. 1978.

Vol. 695: Measure Theory Applications to Stochastic Analys Proceedings, 1977. Edited by G. Kallianpur and D. Kölzow. 261 pages. 1978.

Vol. 696: P. J. Feinsilver, Special Functions, Probability Semigrou and Hamiltonian Flows. VI, 112 pages. 1978.

Vol. 697: Topics in Algebra, Proceedings, 1978. Edited by M. F. Ne man. XI, 229 pages. 1978.

Vol. 698: E. Grosswald, Bessel Polynomials. XIV, 182 pages. 19

Vol. 699: R. E. Greene and H.-H. Wu, Function Theory on Manifo Which Possess a Pole. III, 215 pages. 1979.

Vol. 700: Module Theory, Proceedings, 1977. Edited by C. Faith and S. Wiegand. X, 239 pages. 1979.

Vol. 701: Functional Analysis Methods in Numerical Analysis, Proceedings, 1977. Edited by M. Zuhair Nashed. VII, 333 pages. 1979.

Vol. 702: Yuri N. Bibikov, Local Theory of Nonlinear Analytic Ordinary Differential Equations. IX, 147 pages. 1979.

Vol. 703: Equadiff IV, Proceedings, 1977. Edited by J. Fábera. XIX, 441 pages. 1979.

Vol. 704: Computing Methods in Applied Sciences and Engineering, 1977, I. Proceedings, 1977. Edited by R. Glowinski and J. L. Lions. VI, 391 pages. 1979.

Vol. 705: O. Forster und K. Knorr, Konstruktion verseller Familien kompakter komplexer Räume. VII, 141 Seiten. 1979.

Vol. 706: Probability Measures on Groups, Proceedings, 1978. Edited by H. Heyer. XIII, 348 pages. 1979.

Vol. 707: R. Zielke, Discontinuous Čebyšev Systems. VI, 111 pages. 1979.

Vol. 708: J. P. Jouanolou, Equations de Pfaff algébriques. V, 255 pages. 1979.

Vol. 709: Probability in Banach Spaces II. Proceedings, 1978. Edited by A. Beck. V, 205 pages. 1979.

Vol. 710: Séminaire Bourbaki vol. 1977/78, Exposés 507–524. IV, 328 pages. 1979.

Vol. 711: Asymptotic Analysis. Edited by F. Verhulst. V, 240 pages. 1979.

Vol. 712: Equations Différentielles et Systèmes de Pfaff dans le Champ Complexe. Edité par R. Gérard et J.-P. Ramis. V, 364 pages. 1979.

Vol. 713: Séminaire de Théorie du Potentiel, Paris No. 4. Edité par F. Hirsch et G. Mokobodzki. VII, 281 pages. 1979.

Vol. 714: J. Jacod, Calcul Stochastique et Problèmes de Martingales. X, 539 pages. 1979.

Vol. 715: Inder Bir S. Passi, Group Rings and Their Augmentation Ideals. VI, 137 pages. 1979.

Vol. 716: M. A. Scheunert, The Theory of Lie Superalgebras. X, 271 pages. 1979.

Vol. 717: Grosser, Bidualräume und Vervollständigungen von Banachmoduln. III, 209 pages. 1979.

Vol. 718: J. Ferrante and C. W. Rackoff, The Computational Complexity of Logical Theories. X, 243 pages. 1979.

Vol. 719: Categorial Topology, Proceedings, 1978. Edited by H. Herrlich and G. Preuß. XII, 420 pages. 1979.

Vol. 720: E. Dubinsky, The Structure of Nuclear Fréchet Spaces. V, 187 pages. 1979.

Vol. 721: Séminaire de Probabilités XIII. Proceedings, Strasbourg, 1977/78. Edité par C. Dellacherie, P. A. Meyer et M. Weil. VII, 647 pages. 1979.

Vol. 722: Topology of Low-Dimensional Manifolds. Proceedings, 1977. Edited by R. Fenn. VI, 154 pages. 1979.

Vol. 723: W. Brandal, Commutative Rings whose Finitely Generated Modules Decompose. II, 116 pages. 1979.

Vol. 724: D. Griffeath, Additive and Cancellative Interacting Particle Systems. V, 108 pages. 1979.

Vol. 725: Algèbres d'Opérateurs. Proceedings, 1978. Edité par P. de la Harpe. VII, 309 pages. 1979.

Vol. 726: Y.-C. Wong, Schwartz Spaces, Nuclear Spaces and Tensor Products. VI, 418 pages. 1979.

Vol. 727: Y. Saito, Spectral Representations for Schrödinger Operators With Long-Range Potentials. V, 149 pages. 1979.

Vol. 728: Non-Commutative Harmonic Analysis. Proceedings, 1978. Edited by J. Carmona and M. Vergne. V, 244 pages. 1979.

Vol. 729: Ergodic Theory. Proceedings, 1978. Edited by M. Denker and K. Jacobs. XII, 209 pages. 1979.

Vol. 730: Functional Differential Equations and Approximation of Fixed Points. Proceedings, 1978. Edited by H.-O. Peitgen and H.-O. Walther. XV, 503 pages. 1979.

Vol. 731: Y. Nakagami and M. Takesaki, Duality for Crossed Products of von Neumann Algebras. IX, 139 pages. 1979.

Vol. 732: Algebraic Geometry. Proceedings, 1978. Edited by K. Lønsted. IV, 658 pages. 1979.

Vol. 733: F. Bloom, Modern Differential Geometric Techniques in the Theory of Continuous Distributions of Dislocations. XII, 206 pages. 1979.

Vol. 734: Ring Theory, Waterloo, 1978. Proceedings, 1978. Edited by D. Handelman and J. Lawrence. XI, 352 pages. 1979.

Vol. 735: B. Aupetit, Propriétés Spectrales des Algèbres de Banach. XII, 192 pages. 1979.

Vol. 736: E. Behrends, M-Structure and the Banach-Stone Theorem. X, 217 pages. 1979.

Vol. 737: Volterra Equations. Proceedings 1978. Edited by S.-O. Londen and O. J. Staffans. VIII, 314 pages. 1979.

Vol. 738: P. E. Conner, Differentiable Periodic Maps. 2nd edition, IV, 181 pages. 1979.

Vol. 739: Analyse Harmonique sur les Groupes de Lie II. Proceedings, 1976-78. Edited by P. Eymard et al. VI, 646 pages. 1979.

Vol. 740: Séminaire d'Algèbre Paul Dubreil. Proceedings, 1977–78. Edited by M.-P. Malliavin. V, 456 pages. 1979.

Vol. 741: Algebraic Topology, Waterloo 1978. Proceedings. Edited by P. Hoffman and V. Snaith. XI, 655 pages. 1979.

Vol. 742: K. Clancey, Seminormal Operators. VII, 125 pages. 1979.

Vol. 743: Romanian-Finnish Seminar on Complex Analysis. Proceedings, 1976. Edited by C. Andreian Cazacu et al. XVI, 713 pages. 1979.

Vol. 744: I. Reiner and K. W. Roggenkamp, Integral Representations. VIII, 275 pages. 1979.

Vol. 745: D. K. Haley, Equational Compactness in Rings. III, 167 pages. 1979.

Vol. 746: P. Hoffman, τ-Rings and Wreath Product Representations. V, 148 pages. 1979.

Vol. 747: Complex Analysis, Joensuu 1978. Proceedings, 1978. Edited by I. Laine, O. Lehto and T. Sorvali. XV, 450 pages. 1979.

Vol. 748: Combinatorial Mathematics VI. Proceedings, 1978. Edited by A. F. Horadam and W. D. Wallis. IX, 206 pages. 1979.

Vol. 749: V. Girault and P.-A. Raviart, Finite Element Approximation of the Navier-Stokes Equations. VII, 200 pages. 1979.